Disclaimer

The publisher of this book is by no way associated with the National Institute of Standards and Technology (NIST). The NIST did not publish this book. It was published by 50 page publications under the public domain license.

50 Page Publications.

Book Title: Fire Hazard Comparison of Fire-Retarded and Non-Fire-Retarded Products (NBS SP 749)

Book Author: V Babrauskas; Richard H. Harris; Richard G. Gann; Barbara C. Levin; B T. Lee; Richard D. Peacock; M Paabo; W H. Twilley; M F. Yoklavich; H M. Clark;

Book Abstract: A test program was conducted for the Fire Retardant Chemicals Association to quantify the effects of fire retardant chemicals on total fire hazard. Five different types of products, each made from a different type of plastic were used. The products were made up in analogous fire-retardant (FR) and non-retarded variants (NFR).

Citation: NIST SP - 749

Keyword: flame retardants; cone calorimeters; furniture calorimeters; chromatography; plastics; heat release rate; compartment fires; fire tests; smoke production

U.S. DEPARTMENT OF COMMERCE
National Bureau of Standards

NBS Special Publication 749

Fire Hazard Comparison of Fire-Retarded and Non-Fire-Retarded Products

Vytenis Babrauskas, Richard H. Harris, Jr., Richard G. Gann, Barbara C. Levin,
Billy T. Lee, Richard D. Peacock, Maya Paabo, William Twilley,
Margaret F. Yoklavich, and Helene M. Clark

The National Bureau of Standards[1] was established by an act of Congress on March 3, 1901. The Bureau's overall goal is to strengthen and advance the Nation's science and technology and facilitate their effective application for public benefit. To this end, the Bureau conducts research to assure international competitiveness and leadership of U.S. industry, science and technology. NBS work involves development and transfer of measurements, standards and related science and technology, in support of continually improving U.S. productivity, product quality and reliability, innovation and underlying science and engineering. The Bureau's technical work is performed by the National Measurement Laboratory, the National Engineering Laboratory, the Institute for Computer Sciences and Technology, and the Institute for Materials Science and Engineering.

The National Measurement Laboratory

Provides the national system of physical and chemical measurement; coordinates the system with measurement systems of other nations and furnishes essential services leading to accurate and uniform physical and chemical measurement throughout the Nation's scientific community, industry, and commerce; provides advisory and research services to other Government agencies; conducts physical and chemical research; develops, produces, and distributes Standard Reference Materials; provides calibration services; and manages the National Standard Reference Data System. The Laboratory consists of the following centers:

- Basic Standards[2]
- Radiation Research
- Chemical Physics
- Analytical Chemistry

The National Engineering Laboratory

Provides technology and technical services to the public and private sectors to address national needs and to solve national problems; conducts research in engineering and applied science in support of these efforts; builds and maintains competence in the necessary disciplines required to carry out this research and technical service; develops engineering data and measurement capabilities; provides engineering measurement traceability services; develops test methods and proposes engineering standards and code changes; develops and proposes new engineering practices; and develops and improves mechanisms to transfer results of its research to the ultimate user. The Laboratory consists of the following centers:

- Computing and Applied Mathematics
- Electronics and Electrical Engineering[2]
- Manufacturing Engineering
- Building Technology
- Fire Research
- Chemical Engineering[3]

The Institute for Computer Sciences and Technology

Conducts research and provides scientific and technical services to aid Federal agencies in the selection, acquisition, application, and use of computer technology to improve effectiveness and economy in Government operations in accordance with Public Law 89-306 (40 U.S.C. 759), relevant Executive Orders, and other directives; carries out this mission by managing the Federal Information Processing Standards Program, developing Federal ADP standards guidelines, and managing Federal participation in ADP voluntary standardization activities; provides scientific and technological advisory services and assistance to Federal agencies; and provides the technical foundation for computer-related policies of the Federal Government. The Institute consists of the following divisions:

- Information Systems Engineering
- Systems and Software Technology
- Computer Security
- Systems and Network Architecture
- Advanced Systems

The Institute for Materials Science and Engineering

Conducts research and provides measurements, data, standards, reference materials, quantitative understanding and other technical information fundamental to the processing, structure, properties and performance of materials; addresses the scientific basis for new advanced materials technologies; plans research around cross-cutting scientific themes such as nondestructive evaluation and phase diagram development; oversees Bureau-wide technical programs in nuclear reactor radiation research and nondestructive evaluation; and broadly disseminates generic technical information resulting from its programs. The Institute consists of the following divisions:

- Ceramics
- Fracture and Deformation[3]
- Polymers
- Metallurgy
- Reactor Radiation

[1]Headquarters and Laboratories at Gaithersburg, MD, unless otherwise noted; mailing address Gaithersburg, MD 20899.
[2]Some divisions within the center are located at Boulder, CO 80303.
[3] Located at Boulder, CO, with some elements at Gaithersburg, MD.

NBS Special Publication 749

Fire Hazard Comparison of Fire-Retarded and Non-Fire-Retarded Products

Vytenis Babrauskas, Richard H. Harris, Jr., Richard G. Gann, Barbara C. Levin, Billy T. Lee, Richard D. Peacock, Maya Paabo, William Twilley, Margaret F. Yoklavich, and Helene M. Clark

Fire Measurement and Research Division
Center for Fire Research
National Bureau of Standards
Gaithersburg, MD 20899

Sponsored by:
Fire Retardant Chemicals Association
Lancaster, PA 17604

July 1988

U.S. Department of Commerce
C. William Verity, Secretary

National Bureau of Standards
Ernest Ambler, Director

Library of Congress
Catalog Card Number: 88-600560
National Bureau of Standards
Special Publication 749
Natl. Bur. Stand. (U.S.),
92 pages (July 1988)
CODEN: XNBSAV

U.S. Government Printing Office
Washington: 1988

For sale by the Superintendent
of Documents,
U.S. Government Printing Office,
Washington, DC 20402

Table of Contents

Acknowledgements and Sponsorship   v

List of Figures   vii

List of Tables   ix

Executive Summary   xi

1. **INTRODUCTION** 1
2. **MATERIALS** 3
3. **EXPERIMENTAL** 7
   3.1 Cone Calorimeter   7
   - *3.1.1 Combustion Techniques   7*
   - *3.1.2 Gas Sampling Through Impingers   7*
   - *3.1.3 Test Conditions   8*
   - *3.1.4 Data Collected   8*
   - *3.1.5 Test Results   9*

   3.2 Furniture Calorimeter   10
   - *3.2.1 Experimental Technique – General   10*
   - *3.2.2 Experimental Technique – Specific Details   11*
   - *3.2.3 Gas Sampling Through Impingers   11*
   - *3.2.4 Furniture Calorimeter Results   12*
   - *3.2.5 Direct Comparison to Cone Calorimeter Results   12*
   - *3.2.6 Use of Predictive Small-Scale/Large-Scale Correlations   13*

   3.3 Small-Scale Toxicity Tests   14
   - *3.3.1 Animals   14*
   - *3.3.2 Small-Scale Toxicity Test System   14*
   - *3.3.3 Gas Analysis   15*
   - *3.3.4 N-Gas Model Prediction   15*
   - *3.3.5 Results and Discussion of the Small-Scale Toxicity Tests   16*

4. **MODEL SIMULATIONS OF THE LARGE-SCALE ROOM/CORRIDOR/ROOM FIRES** 35
   4.1 Product Arrangement A and Room Corridor Configuration A   35
   - *4.1.1 NFR Products   35*
   - *4.1.2 FR Products with Auxiliary Burner   35*
   - *4.1.3 Model Predictions for Configuration A   36*

   4.2 Product Arrangement B and Room/Corridor Configuration B   36

4.3 Worst Case Scenario for Product Arrangement B and Room/Corridor Configuration B 36
       4.3.1 NFR Products 36
       4.3.2 FR Products 37
       4.3.3 Model Predictions With Configuration B 37
   4.4 Use of the Model Predictions in the Design of Large-Scale Tests 38

5. LARGE-SCALE TESTS ... 49
   5.1 Experimental 49
       5.1.1 Room/Corridor/Room Configuration 49
       5.1.2 Room Fire Test Arrangement 49
       5.1.3 Instrumentation 50
   5.2 Test Procedure 51
   5.3 Large-Scale Test Results 52
       5.3.1 Data from Exhaust Stack Instruments 52
       5.3.2 Data from Animal Chambers 52
       5.3.3 Data from Other Observations 53
   5.4 Comparison Between Small-Scale and Large-Scale Toxicity Findings 53
       5.4.1 Species Yields 53
       5.4.2 Fraction of Toxicity Accounted by CO 54
       5.4.3 Combustion Products of Extreme Toxic Potency 55
       5.4.4 Combustion Products of Unusual Toxic Potency 55
   5.5 Comparison Between Computer Model Predictions and Actual Large-Scale Results 55

6. ASSESSMENT OF HAZARD ... 69
   6.1 Times to Untenability 69
   6.2 Amounts of Combustion Products 70

7. SUMMARY ... 73
   7.1 Properties of Individual Products Tested 73
   7.2 Results from Large-scale Evaluations of NFR- and FR-Furnished Rooms 74
   7.3 Relations Between Measurements in Different Test Methods 75

8. CONCLUSION ... 77

Appendix A. Ion Chromatography Procedure 79

Appendix B. Gas Chromatography Procedure 81

Appendix C. Log of Large-Scale Test Observations 83

References 85

Acknowledgements and Sponsorship

J. Newton Breese, Phyllis Martin, and Carmen Davis provided valuable assistance in the large-scale aspects of the testing program.

The study was sponsored by the Fire Retardant Chemicals Association, P.O. Box 3535, Lancaster, PA 17604. The work was under the auspices of FRCA's Hazard Assessment Committee, comprised of the following firms contributing financial and consultative support:

Albright and Wilson Americas, Inc.
ALCOA
Ameribrom, Inc.
Anzon, Inc.
Asarco, Inc.
BorgWarner Chemicals Inc.
Chemical Industrial Association Ltd.
Dow Chemical USA
Ethyl Corporation
Ferro Corporation
Great Lakes Chemical Co.
Laurel Industries, Inc.
M & T Chemicals Inc.
McGean-Rohco Inc.
Occidental Chemical Co.
Solem Industries, Inc.

List of Figures

Figure 1. TV cabinets used in the tests. ...4
Figure 2. Business machine housings used in the tests. ...4
Figure 3. Upholstered chair cushions used in the tests. ...5
Figure 4. Pieces of the cables used in the tests. ...5
Figure 5. Pieces of the circuit boards used in the tests. ...5
Figure 6. Conceptual view of the Cone Calorimeter. ...25
Figure 7. Smoke measurement system used on the Cone Calorimeter. ...25
Figure 8. Impinger gas sampling in the Cone Calorimeter. ...26
Figure 9. Furniture Calorimeter test of business machines and TV cabinets. ...26
Figure 10. Furniture Calorimeter test of chairs. ...26
Figure 11. Furniture Calorimeter test of cables (Z-configuration). ...27
Figure 12. Furniture Calorimeter test of cables (vertical array). ...27
Figure 13. Furniture Calorimeter test of circuit boards. ...28
Figure 14. Impinger gas sampling in the Furniture Calorimeter. ...28
Figure 15. Rate of heat release for TV cabinets measured in the Furniture Calorimeter. ...29
Figure 16. Rate of heat release for business machine housings measured in the Furniture Calorimeter. ...29
Figure 17. Rate of heat release for upholstered chairs measured in the Furniture Calorimeter. ...30
Figure 18. Rate of heat release for cables measured in the Furniture Calorimeter. ...30
Figure 19. Rate of heat release for circuit boards measured in the Furniture Calorimeter. ...31
Figure 20. Schematic of gas analysis system used in the small-scale toxicity test. ...31
Figure 21. Impinger gas sampling in the small-scale toxicity test. ...32
Figure 22. Comparison of materials by their EC_{50}, LC_{50} (30 min.) and LC_{50} (30 min. + 14 days) after flaming decomposition. ...32
Figure 23. Animal weight change following a 30 minute exposure to the flaming decomposition products from a mass loading of 20 mg/l of specimen K. ...33
Figure 24. Arrangement A of test items in the burn room. ...38
Figure 25. Plan view of large-scale test arrangement A. ...38
Figure 26. Mass loss rate for NFR furnishing arrangement A and room-corridor configuration A without the auxiliary burner. ...39
Figure 27. Composite mass loss history for NFR furnishings. ...39
Figure 28. Mass loss rate for FR furnishing arrangement A and room-corridor configuration A with the auxiliary burner. ...40
Figure 29. Plan view of large-scale test arrangement B. ...40
Figure 30. Arrangement B of test items in the burn room. ...41
Figure 31. Mass loss rate for NFR furnishing arrangement B and room-corridor configuration B without auxiliary burner. ...41
Figure 32. Accelerated mass loss rate for NFR furnishing arrangement B and room-corridor configuration B without the auxiliary burner. ...42

Figure 33. Accelerated mass loss rate for NFR furnishing arrangement B and room-corridor configuration B with the auxiliary burner. 42
Figure 34. Mass loss rate for FR furnishing arrangement B and room-corridor configuration B without the auxiliary burner. 43
Figure 35. Mass loss rate for FR furnishing arrangement B and room-corridor configuration B with the auxiliary burner. 43
Figure 36. Model prediction and test result curves for the average upper layer temperature in the burn room/corridor/target room for NFR furnishings without the auxiliary burner. 44
Figure 37. Model prediction and test result curves for carbon monoxide in the burn room/corridor/target room for NFR furnishings without the auxiliary burner. 44
Figure 38. Model prediction and test result curves for the average upper layer temperature in the burn room/corridor/target room for NFR furnishings with the auxiliary burner. 45
Figure 39. Model prediction and test result curves for carbon monoxide in the burn room/corridor/target room for NFR furnishings with the auxiliary burner. 45
Figure 40. Model prediction and test result curves for average upper layer temperature in the burn room/corridor/target room for FR furnishings without the auxiliary burner. 46
Figure 41. Model prediction and test result curves for carbon monoxide in the burn room/corridor/target room for FR furnishings without the auxiliary burner. 46
Figure 42. Model prediction and test result curves for the average upper layer temperature in the burn room/corridor/target room for FR furnishings with the auxiliary burner. 47
Figure 43. Model prediction and test result curves for carbon monoxide in the burn room/corridor/target room for FR furnishings with the auxiliary burner. 47
Figure 44. Impinger gas sampling in the large-scale burn room. 64
Figure 45. Impinger gas sampling in the large-scale animal chamber. 64
Figure 46. Heat release rates in the large-scale room/corridor/room tests without the auxiliary burner. 65
Figure 47. Heat release rates in the large-scale room/corridor/room tests with the auxiliary burner. 65
Figure 48. Mass flow rates of CO in the large-scale room/corridor/room tests without the auxiliary burner. 66
Figure 49. Mass flow rate of CO in the large-scale room/corridor/room tests with the auxiliary burner. 66
Figure 50. Mass flow rate of CO_2 in the large-scale room/corridor/room tests without the auxiliary burner. 67
Figure 51. Mass flow rate of CO_2 in the large-scale room/corridor/room tests with the auxiliary burner. 67
Figure 52. Smoke flow rates in the large-scale room/corridor/room tests without the auxiliary burner. 68
Figure 53. Smoke flow rates in the large-scale room/corridor/room tests with the auxiliary burner. 68

List of Tables

Table 1	Cone Calorimeter Data Summary—30 kW/m² Irradiance Tests	18
Table 2	Cone Calorimeter Data Summary—Test Average Data at 30 kW/m² Irradiance	19
Table 3	Cone Calorimeter Data Summary—100 kW/m² Irradiance Tests	19
Table 4	Cone Calorimeter Data Summary—Test Average Data at 100 kW/m² Irradiance	20
Table 5	Irradiance Threshold Limit for Foam S with Nylon Fabric Cover	20
Table 6	Summary of Furniture Calorimeter Test Conditions	21
Table 7	Summary of Furniture Calorimeter Results	21
Table 8	Comparison of Cone Calorimeter Versus Furniture Calorimeter Data	22
Table 9	N-gas Model Toxicity Procedure	22
Table 10	Chemical and Toxicological Results from Materials Decomposed in the Flaming Mode	23
Table 11	Summary of LC_{50} Values and 4-Gas Model Prediction Values Determined in the Small-Scale Toxicity Tests	24
Table 12	Dimensions of Room/Corridor/Room Large-Scale Test Facility	56
Table 13	Construction of Large-Scale Test Facility	56
Table 14	Location of Instrumentation	56
Table 15	Ratio of Gases in Animal Chambers to Target Room for Specified Times	57
Table 16	Animal Exposure Chamber Filling Times and Estimated CO When Closed	58
Table 17	Chemical and Toxicological Results in Animal Exposure Chambers Filled During Large-Scale Room Burns	59
Table 18	Comparison of Peak Times and Concentration Areas for CO, HCl, HBr and HCN in Large-Scale Tests (Burn Room)	59
Table 19	Specimen Mass Consumed in Large-Scale Tests	60
Table 20	Comparison Between Yields of Toxic Species in the Different Devices	60
Table 21	Fraction of Total Toxicity Accounted for by CO (NBS Combustion Toxicity Test)	61
Table 22	Fraction of Total Toxicity Accounted for by CO (Large-Scale)	61
Table 23	Times to Reach Untenable Conditions in Large-Scale Tests	62
Table 24	Comparison of Total Heat Release from Large-Scale Fires with Furniture Calorimeter and Cone Calorimeter Calculated Values	62
Table 25	Comparison of Smoke from Large-Scale Fires with Furniture Calorimeter and Cone Calorimeter Calculated Values	62
Table 26	Comparison of Average CO from Large-Scale Fires with Furniture Calorimeter, Cone Calorimeter, and Toxicity Test Calculated Values	63
Table 27	Total Toxicity Results in Large-Scale Tests	63

Executive Summary

BACKGROUND

Fire retardants (FR) are most frequently added to plastics in order to reduce their burning rate. Historically, this has meant increasing a material's resistance to a variety of Bunsen burner type exposures. These simple, visual tests have resulted in the modification of many of the most obviously flammable materials, and an accordant increase in fire safety.

In recent years, the public has perceived a broader concept of product fire safety. This advanced view includes three contributions to fire hazard:

- Rate of fire growth. This is measured as the rate of heat release from the burning material and the resulting increased temperatures near and away from the fire.
- Smoke obscuration. The time-variant yield of soot and the nature of that soot affect both the spread of alarm and the ability of alerted people to escape.
- Smoke toxicity. Inhalation of the fire products can result in a variety of ill effects ranging from disorientation to death.

This new understanding has led to reconsideration of the overall fire safety of commercial building and furnishing materials. For fire-retardant products in particular, the question has been raised as to (a) whether the additives effect a trade-off between decreased burning and increased emission of toxic gas species (smoke) and (b) whether there is a net safety benefit from the use of fire retardants.

During the past several years, the National Bureau of Standards' Center for Fire Research (NBS-CFR) has been developing the methodology to determine the overall fire hazard of commercial products. This includes the use of advanced bench-scale measurement methods, computer modeling of fires, and confirmatory large-scale tests.

The Fire Retardant Chemicals Association (FRCA) therefore asked NBS-CFR to investigate the fire hazard of a wide array of fire-retardant (FR) containing products relative to non fire-retarded (NFR), but otherwise substantially identical, products. The two central issues to be explored were:

(1) For today's most commonly used FR/polymer systems, is the overall fire hazard reduced, when compared to similar non-fire retarded (NFR) items?

(2) Since both the commercially popular FR chemicals and the base polymer formulations can be expected to change in the future, can appropriate bench-scale test methodologies be validated which would allow future testing to be quick and simple?

APPROACH

To answer these questions, a wide-ranging experimental program was formulated. Five representatives of commonly used plastic products were especially manufactured (using commercial formulations) for this program, each in an NFR and a FR version. These were:

- polystyrene television cabinet
- polyphenylene oxide business machine housing
- polyurethane foam-padded upholstered chair
- electrical cable with polyethylene wire insulation and rubber jacketing
- polyester/glass electric circuit board

The test program involved bench-scale tests of the individual products in the Cone Calorimeter, where rate of heat release (the best quantitative measure of burning rate), ignitability, rate of smoke production, and the rates of production of various toxic combustion gases were determined. Bench-scale tests were also conducted on each product using the NBS cup furnace combustion toxicity apparatus. Entire articles were then tested in the Furniture Calorimeter, which allows measurements similar to those taken in the Cone Calorimeter, but at full-scale and burning in an open environment. The final proof testing took place in a large-scale burn room/corridor/target room facility, where instead of examining the specimens singly, an arrangement which included either all the NFR or all the FR products was used.

Before this final testing could be done, assurance was obtained that a useful fire scenario had been developed. For the first time, computer fire modeling was used to determine that the NFR room had enough combustibles to assure a rapid, high intensity fire, and that the concentration and flow of the combustion gases were such that useful toxicological quantification by means of a bioassay (animal exposure) could be done. It was also important to assure that, in the FR case, the fire would not simply die out because the initial item ignited failed to spread flame. (Such an outcome would not provide assurance that, had the ignition source been just slightly bigger, the fire would have spread and much more hazardous conditions resulted.)

THE LARGE-SCALE RESULTS

The impact of FR materials on the survivability of the building occupants was assessed in two ways: (1) Comparing the *time* to untenability in the burn room; this is applicable to the occupants of the burn room. (2) Comparing the total *production* of heat, toxic gases, and smoke from the fire; this is applicable to occupants of the building remote from the room of fire origin.

The time to untenability is judged by the time that is available to the occupants before the earlier of (a) room flashover, or (b) untenability due to toxic gas production occurs. For the FR tests, the average available escape time was more than 15-fold greater than for the occupants of the NFR room. With regard to the production of combustion products,

- The amount of material consumed in the fire for the FR tests was less than half the amount lost in the NFR tests.
- The FR tests indicated an amount of heat released from the fire which was 1/4 that released by the NFR tests.
- The total quantities of toxic gases produced in the room fire tests, expressed in "CO equivalents," were 1/3 for the FR products, compared to the NFR ones.
- The production of smoke was not significantly different between the room fire tests using NFR products and those with FR products.

Thus, in these tests, the fire retardant additives did decrease the overall fire hazard of their host products.

The above conclusions are specifically pertinent only to the materials actually examined. Thus, while it has been demonstrated that very significantly enhanced fire performance can be obtained with fire-retarded products, such improvements are by no means to be automatically expected from all fire-retarded products. Instead, it will still be necessary to test and evaluate proposed new systems individually. However, these tests do show that the proper selection of fire retardants can markedly improve the fire safety of specific products.

THE RESULTS OF SMALLER-SCALE TESTS

Prediction of room fire behavior from smaller-scale test data is a relatively recent area of fire research. A full capability requires that both the burning rate (heat release rate or mass loss rate) and the relevant yields of combustion products could be predicted. Attempts to calculate burning rates are a new area of endeavor; and there are a number of physical phenomena which cannot be adequately represented by the smaller-scale test itself, but must be addressed using special data analysis techniques or by empirical correction factors.

Without such detailed techniques in hand, it was still desired to find out if the Cone Calorimeter rate of heat release measurements of products could indicate the level of real-scale improvement which the FR additives could effect. This was found to be true for 4 of the 5 product categories tested. The Cone Calorimeter results for the electric cables did not show adequate correlation.

Small-scale toxicity tests were used by themselves to answer two questions:

- Do the combustion products from the specimens, when burned under flaming conditions, show extreme toxicity?
- Could the animal mortality be predicted from the analyses of the concentrations of the principal toxic fire gases and knowledge of their toxicological interactions? (That is, is there no unusual toxicity?)

The results showed that none of the test specimens produced smoke of extreme toxicity. The smoke from both the FR and the NFR products was similar in potency and comparable to the potency of smoke produced by materials commonly found in buildings. For

7 out of the 10 specimens studied, the results can be predicted from the 4-gas model; the 3 exceptions are the FR polystyrene TV cabinet, the FR ethylene vinyl acetate wire insulation, and the FR polyester circuit board. For those three cases, unidentified agents made a small contribution to the specimen's toxicity. Identification of these agents was not pursued because they were minor.

The large-scale studies generally confirmed the small-scale toxicity results. Although the uncertainties were larger and individual products were not isolated in large-scale testing, the large-scale data do show that neither extreme nor unusual toxicity was found.

It was further found that the fraction of total toxicity accounted for by CO in the small-scale toxicity test and the fraction measured in the large-scale room/corridor/room tests were similar. This encourages the use of the NBS combustion toxicity test to characterize the gross toxicity of combustion gases.

The gas species yields at both scales were also compared directly. CO_2 and the acid gas (HCl, HBr, HCN) yields showed agreement in all cases where their concentrations were well above the background. For CO, there was substantive disagreement, with the NBS combustion toxicity apparatus and, especially, the Cone Calorimeter giving some values which were substantially lower than those observed in the Furniture Calorimeter. The values derived from large-scale tests were, in turn, not very well predicted by the Furniture Calorimeter data, even though the scale was similar.

In all cases, the ratios between the smoke yields for the NFR and the FR products were close in the Cone Calorimeter and in the Furniture Calorimeter. However, the ratio between the numerical values in the Furniture Calorimeter and those in the Cone Calorimeter was typically 1:2. The exception to this enhanced smoke yield in the smaller-scale test came from the polystyrene TV cabinets. There, both the NFR and the FR versions showed greater smoke yields in the Furniture Calorimeter than in the Cone Calorimeter.

Thus, the answer to the second major question—Can bench-scale test methodologies suffice for evaluating future products?—is mixed. In some cases, the smaller-scale tests provide indicative data; in others, they do not. This finding is consistent with the current development of predictive methods for the Cone Calorimeter or the Furniture Calorimeter on a product-class basis. It remains for the rapidly-progressing fire hazard modeling to provide a more universal relationship between bench-scale measurements of fire properties and the fire threats from commercial products.

Introduction

FIRE retardants are most frequently added to plastics in order to reduce their burning rate. Historically, this has meant increasing a material's resistance to a variety of Bunsen burner type exposures. These simple, visual tests have resulted in the modification of many of the most obviously flammable materials, and an accordant increase in fire safety.

In recent years, the public has perceived a broader concept of product fire safety. This advanced view includes three contributions to fire hazard:

- Rate of fire growth. This is measured as the rate of heat release from the burning material and the resulting increased temperatures near and away from the fire.
- Smoke obscuration. The time-variant yield of soot and the nature of that soot affect both the spread of alarm and the ability of alerted people to escape.
- Smoke toxicity. Inhalation of the fire products can result in a variety of ill effects ranging from disorientation to death.

This new understanding has led to reconsideration of the overall fire safety of commercial building and furnishing materials. For fire-retardant products in particular, the question has been raised as to (a) whether the additives effect a trade-off between decreased burning and increased emission of toxic gas species (smoke) and (b) whether there is a net safety benefit from the use of fire retardant.

During the past several years, the National Bureau of Standards' Center for Fire Research (NBS-CFR) has been developing the methodology to determine the overall fire hazard of commercial products. [*Fire hazard* refers to the level of threat presented by a fire to people and structures. For example, the fire hazard to the sleeping occupants of a wooden shack from an indoor usage of gasoline for heating purposes might be high: no one would escape and the entire structure would be destroyed. *Fire risk*, by contrast, weighs the possible hazards by the likelihood they will occur. Thus, the risk from the fire above might be quite low because such a situation is rare.] The assessment of fire hazard involves the use of advanced bench-scale measurement methods, computer modeling of fires and their impact, and confirmatory large-scale tests. The Fire Retardant Chemicals Association (FRCA) therefore asked NBS-CFR to investigate the fire hazard of a wide array of fire-retardant (FR) containing products relative to non fire-retarded (NFR), but otherwise similar, products. The central question to be explored was: Can these FR products be considered to provide a higher all-around degree of fire safety than their NFR counterparts? This is the first time that such a global test would be applied to such a wide range of products.

Five types of products were selected, each chosen to represent the potential use of a different type of polymer and a different FR formulation. In each case, an NFR control was prepared, similar to the FR product, but containing no FR agent. The commodities were to be studied using the following:

- Cone Calorimeter
- Furniture Calorimeter
- Small-Scale N-Gas Model Toxicity Procedure
- Fire hazard modeling (FAST)
- Multi-room large-scale fire tests

In the Cone Calorimeter [1],* bench-scale specimens cut from the test products were evaluated for several fire properties: effective heat of combustion, and rates of heat release, smoke, and several different gas species production. In the Furniture Calorimeter

---

*Numbers in brackets refer to literature references listed in Section 9 at the end of this report.

[2], real-scale specimens were tested for the same variables.

In the Small-Scale N-Gas Model Toxicity Procedure [3,4], the smoke from each burning product was evaluated for its toxic potency and any unusual toxic gas generation using a modified animal-exposure methodology.

These small-scale data were then combined as needed and used as input to FAST, the fire growth and smoke transport program in the prototype version of HAZARD-I [5]. The resulting predictions of fire development were used to guide the design of the multi-room tests.

Finally, in large-scale fire experiments, all of the specimens (either NFR or FR) were grouped and burned to determine heat release rate, smoke, toxic gas production, and actual animal toxicity. These tests served as the ultimate demonstration of actual differences between the fire hazard performance of the NFR and the FR products.

Before proceeding to the detailed description of the study, two items of clarification are in order. First, when discussing products of combustion, we will distinguish between *production* and *yield*. Production is the total amount of a substance evolved during the duration of a test; yield is that production normalized by the mass of specimen lost (burned) during the period. Results from disparate test devices are most properly compared on a yield basis, while hazard assessment usually requires knowledge of the production involved.

Second is the definition of similar or different results. The individual test measurements to be presented typically have good repeatabilities, on the order of 10% or better. Therefore, the assessment of similarity is simply a matter of comparing the differences in values with the uncertainties in their measurements. By contrast, a single scenario will be developed for large-scale testing, and the results will be compared in the context of that scenario. The scenario chosen, while well-grounded, can by no means be unique. Indeed, a different scenario choice could seriously influence the conclusions in a manner that can not be determined in this work.

Materials

FIVE different types of products were evaluated in this study. Each was characterized with and without appropriate fire retardants. Fire-retardant product formulations were chosen to represent ones which are commercially available and in common use, but which were also anticipated to represent high quality performance. The items studied were as follows:

(1) TV Cabinet housing. The TV cabinets were moldings of an external cabinet only; no internal working parts were used. The average thickness of the molded material was 3.0 mm. The assembled cabinets are shown in Figure 1.

  Sample H (NFR) — high impact polystyrene base formulation;

  Sample G (FR) — the same base formulation with decabromodiphenyl oxide (12% by weight) and antimony oxide (4%)

(2) Business machine housing. The business machine housings were also moldings of an external cabinet only and were of similar physical appearance as the TV cabinet housings; again, no internal working parts were used. The average thickness of the molded material was 3.0 mm. The assembled housings are shown in Figure 2.

  Sample F (NFR) — poly(2,6-dimethyl 1,4 phenylene) oxide; also includes polystyrene, polybutadiene, polyethylene, mineral oil, and stabilizer additives.

  Sample A (FR) — the same base formulation, with a triaryl phosphate ester based flame-retardant (to give 1% P by weight).

(3) Upholstered chairs. The upholstered chairs were constructed of only two combustible materials: flexible polyurethane padding foam, and a cover fabric. Instead of a conventional frame, the chairs used a steel mock-up frame. The assembled chairs are shown in Figure 3.

  Sample T (NFR) — The density of this foam was 25 kg/m^3.

  Sample S (FR) — This foam contained an organic chlorinated phosphate, and organic brominated retardant and 35% alumina trihydrate. The loadings represented an elemental content of 4.75% Br, 2.6% Cl, 0.32% P, and 10.0% Al. The density of this foam was 64 kg/m^3.

  The same nylon fabric (250 kg/m^2) was used as a cover for both samples. Since the cover fabric was not varied, it was not evaluated in certain of the bench-scale tests.

(4) Cable array. Each electric cable contained five copper wires, each 14 AWG (1.63 mm dia.). The outside diameter of each insulated wire was 3.30 mm. The overall, outside diameter of the complete jacketed cable was 12.7 mm. Pieces of the cable approximately 250 mm long are shown in Figure 4.

  Sample D (NFR) — wire insulation made of crosslinked ethylene vinyl acetate copolymer, with clay (18.9 parts per 100 resin), antioxidant (2 parts), processing aid (1 part), and catalyst (1.5 parts). Covered with a black outside jacket made of chlorosulfonated polyethylene containing Sb_2O_3. Elemental contents were 12.2% Cl and 2% Sb.

  Sample K (FR) — wire insulation made of polyethylene cross-linked with ethylene vinyl acetate, with clay (28 parts), chlorinated cycloaliphatic fire retardant (38 parts), Sb_2O_3 (18.9 parts), antioxidant (2 parts), processing aid (1 part), and catalyst (1.5 parts). Outside jacket identical to that for the NFR specimen.

  The outer jackets, being the same in both instances, were not evaluated in detail in certain of the bench-scale tests.

(5) Laminated circuit board This material was intended to be representative of glass/polyester electric circuit boards. It contained, however, no copper traces and no electrical components. The thickness of the board was 6.4 mm, and it was supplied in large sheets, which were cut up for use. Pieces of each circuit board are shown in Figure 5.

Sample C (NFR) — polyester resin (38% by weight), with $CaCO_3$ filler (44% by weight), and fiberglass reinforcement (18%).

Sample L (FR) — polyester resin (39%), with decabromodiphenyl oxide (10%), Sb_2O_3 (3%), $Al_2O_3 \cdot 3H_2O$ (30%), and fiberglass reinforcement (18%).

Figure 1. TV cabinets used in the tests.

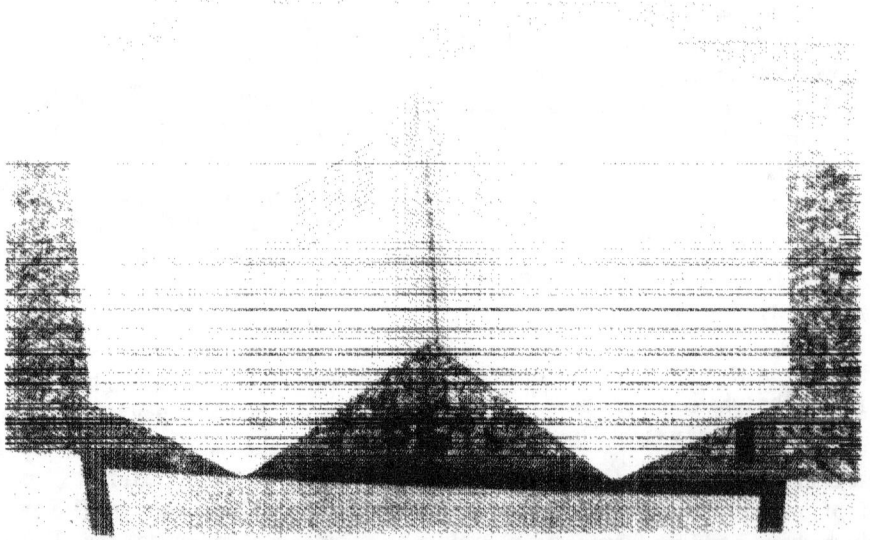

Figure 2. Business machine housings used in the tests.

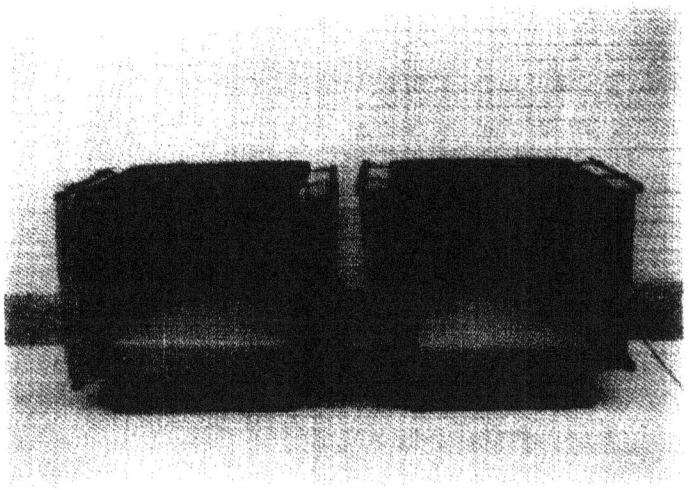

Figure 3. Upholstered chair cushions used in the tests.

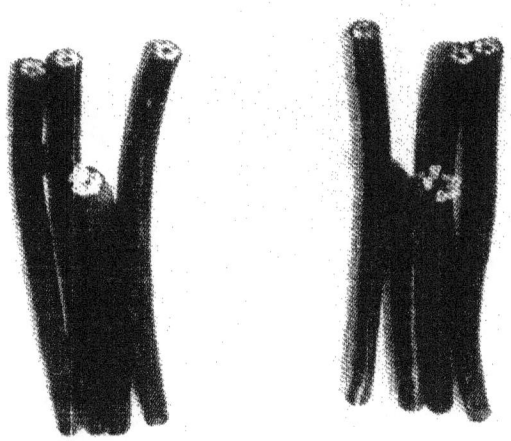

Figure 4. Pieces of the cables used in the tests.

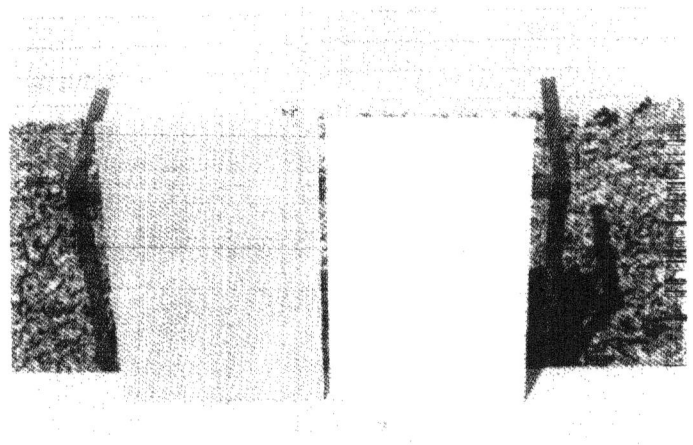

Figure 5. Pieces of the circuit boards used in the tests.

Experimental

3.1 CONE CALORIMETER

3.1.1 COMBUSTION TECHNIQUES

THE Cone Calorimeter (Figure 6) was initially presented in 1982 as an improved technique for measuring rate of heat release on bench-scale specimens [1]. Its operation involves an application of the oxygen consumption principle. Earlier instruments for measuring rates of heat release were based on either a concept of a well-insulated box, which led to some very serious measurement errors, or on substitution type schemes, which were cumbersome and difficult to operate.

The oxygen consumption principle [7] states that for most combustibles there is a unique constant, 13.1 MJ/kg O_2, relating the amount of heat released during a combustion reaction and the amount of oxygen consumed from the air. Thus, using this principle it is only necessary to measure the concentration of oxygen in the combustion stream, along with the flow rate.

In the Cone Calorimeter, specimens of a material or product to be tested are cut into a 100 mm by 100 mm size. The thickness depends on the type of product tested, and can range from 6 to 50 mm. The specimen edges are protected from burning, and the specimen can be oriented either horizontally or vertically. In the FRCA program, all specimens were tested horizontally, face up.

The specimen is heated by an electric heater in the shape of a truncated cone, hence, the name Cone Calorimeter. The irradiance to the specimen can be set to any desired value from zero to 110 kW/m². If required, external ignition of the specimen is provided by an electric spark. Since a uniform, controlled irradiance is provided, the ignition times themselves, as measured, constitute a suitable test for ignitability.

The specimen is mounted on a load cell and its mass, along with all other instrument data, is reorded every 5 s.

A few years ago, when shortcomings of existing smoke measuring tests were becoming evident, a smoke measuring system (Figure 7) was evolved for the Cone Calorimeter. This comprises a He-Ne laser beam projected across the exhaust duct. The monochromatic light is monitored by a solid-state detector. A second detector serves as a reference, to guard against effects of drift and of laser power fluctuations. The optical system is designed to be self-purging, and does not use optical windows. Further details of the smoke measuring system have been given in a recent publication [8].

An overview of the design features, along with the specifications and limitations of the Cone Calorimeter has recently been published [9].

3.1.2 GAS SAMPLING THROUGH IMPINGERS

Since the Cone Calorimeter represents a well-controlled combustion environment to which specimens can be exposed, it became advantageous to also use it for the measurement of gaseous species yields. Thus, fixed gas analyzers for O_2, CO, CO_2, total hydrocarbons, and H_2O vapor have been implemented. For the testing of FRCA materials, however, it was also necessary to characterize the yields of certain acid gases—HCl, HBr, and HCN. For these, commercial on-line gas analyzers were not available, thus a batch sampling technique needed to be employed.

A portion of the gaseous products and soot in the main exhaust duct of the Cone Calorimeter was collected by replacing the soot collection filter with a gas sampling apparatus. Figure 8 shows a diagram of the gas sampling apparatus. The gaseous products were collected in tared 250 ml glass impinger bottles containing approximately 125 ml of 5 mM KOH. To maximize the collection efficiency, two impingers were used in series, separated by a 45 mm diameter PTFE filter (0.45 μm nominal porosity) to collect the sample from the exhaust stream. The second impinger served to trap any gases that might break through the PTFE filter. Because only a small amount of breakthrough was detected, the second impinger was not used later in the program. The flow of gases through the impingers was controlled by the existing mass flow controller used for soot collection. The ratio of gases collected to gases exhausted was nominally 1:1000; however, the exact value for each test was recorded and used in computations. Sample collection started when the specimen was placed on the load cell of the calorimeter; the collection was stopped when the specimen stopped burning and no more smoke was being evolved.

After the tests were completed, the impingers were weighed and the contents transferred to plastic containers. Prior to analysis, the filter containing the soot was placed into the solution in the first impinger. The pH of the unknown solutions was determined to make sure that it did not drop below the optimum range for the ion chromatographic analysis. The samples were then analyzed for the expected anion(s) by the procedure described in Appendix A.

3.1.3 TEST CONDITIONS

The tests were conducted according to the ASTM (American Society for Testing and Materials) method P 190 [10]. In addition, besides specifying the specimen orientation (horizontal, face-up), and the use of spark ignition, to describe fully the test conditions requires specifying the test irradiance and any special specimen preparation techniques.

Three different irradiance values were used in the test program:

- 10 kW/m^2 (with spark ignition)
- 30 kW/m^2 (with spark ignition)
- 100 kW/m^2 (without spark ignition)

The lowest irradiance value, 10 kW/m^2, would not be expected to cause ignition in most specimens; specimens which are *unusually* ignition-prone would, however, be ignited. The intermediate value, 30 kW/m^2, represents a heating value that can readily be imposed by one free-burning item on another. Many bench-scale tests also impose heating levels in the range of 30 kW/m^2 on test specimens. The high value, 100 kW/m^2, was selected to monitor if any unusual combustion phenomena become manifest at high irradiances, which are not evident during normal testing. Such high irradiances are typical of the upper ranges of heating values imposed on materials in a fully involved (flashed over) room fire.

All specimens, as the final step of specimen preparation, were wrapped in aluminum foil on the bottom and sides. Otherwise, the TV cabinet housings, the business machine housings and the circuit board specimens were prepared by cutting 100 mm by 100 mm coupons from the full-scale articles supplied, and were not otherwise treated. The upholstered chair specimens were prepared by cutting foam specimens slightly smaller than 100 mm by 100 mm by 50 mm thick. The cover fabric was cut into 200 mm by 200 mm samples, and from each corner a 50 mm by 50 mm square was removed. The foam was then covered with the fabric over its top and 4 sides. The fabric was stapled to the foam along the bottom edge of all sides. Since the FR product, in this case, was an inner, rather than outer, layer, the furniture foam samples were also tested alone, without the use of the cover fabric. The test cables were prepared by cutting them into 100 mm lengths and laying up a single row, side-by-side, in the aluminum foil. Two supplementary configurations on the cables were also tested, where (1) the outer rubber jacket was removed, and only the individual wires (copper + wire insulation) were tested, and (2) the outer rubber jacket alone was tested.

The specimen thicknesses, as tested, were:

- TV cabinet housing—3 mm
- Business machine housing—3 mm
- Upholstered chair—50 mm
- Cable array—13 mm
- Laminated circuit board—6 mm

The thicknesses of the TV cabinet and business machine housings were thinner than the normal test thickness of 6 mm. For such cases, the test method prescribes that the specimen should be tested in conjunction with its end-use substrate. Since, in the present test series, the exact arrangement of the electronic equipment in the TV cabinet and business machine housings could not be stated, a back face condition comprising only the normal refractory fiber blanket was used.

3.1.4 DATA COLLECTED

The data to be derived from the bench-scale tests in the Cone Calorimeter constitute a very large set, and can be analyzed in a multitude of ways. The most important variables which are presented include the following:

- peak rate of heat release (kW/m²)
- rates of heat release averaged over various time periods, starting with the time of ignition (kW/m²)
- effective heat of combustion (MJ/kg). This will be less than the oxygen-bomb value of the heat of combustion, since the combustion is incomplete.
- percent specimen mass lost (%)
- time to ignition (s)
- average smoke obscuration (m²/kg). Smoke production from a material has the rational units of m², representing the extinction cross-section of the smoke. This is normalized by the amount of specimen mass lost (kg).
- average yields of each of the measured gas species (kg/kg)

The total number of replicate tests run varied between two and six, with the higher number generally being run on specimens showing wider data scatter.

3.1.5 TEST RESULTS

The Cone Calorimeter results are summarized in Tables 1 through 5. For the 10 kW/m² irradiance, even though the spark ignitor was used, none of the test specimens ignited. Therefore, detailed data at 10 kW/m² are not tabulated and will not be further discussed. Detailed data sets (Tables 1 to 4) were collected at 30 kW/m² and at 100 kW/m². The values reported, unless otherwise noted, are the average of two to six tests. Measures of data uncertainty are given in Tables 1 and 2. The discussion below focuses mainly on the 30 kW/m² results; where the findings at 100 kW/m² are substantively different, the 100 kW/m² data will also be discussed.

At this point, only the raw data from the bench-scale Cone Calorimeter tests are first presented. Thus, the reader is cautioned that differences in performance noted among the specimens refer only to the bench-scale performance. The comparison to large-scale findings, and discussions of the general usefulness of bench-scale methods for predicting large-scale fire results are topics taken up in the subsequent sections of this report. (See Sections 7 and 8.)

3.1.5.1 TV cabinet housings

For the TV cabinet housings, specimens H and G showed roughly similar fractions of mass burned. The peak rate of heat release of the FR specimen G was only 35% of the value for the un-retarded specimen H. This is a significant improvement. Specimen G only released half as much total heat as did H, while its effective heat of combustion was some 2.5 times smaller. The ignition time of the FR specimen, however, was somewhat shorter than for the NFR one. The smoke yield *per unit mass loss* of specimen, however, was about twice for the FR specimen as for the NFR one. The yield of CO was about 7-fold greater for the FR specimen, compared to the NFR one, at 30 kW/m² irradiance. However, since at 100 kW/m² the yields of CO became essentially the same, no special significance is attributed to the finding. The CO_2 yield for the FR specimen was some 3 times less than for the NFR one under either irradiance condition. Since the amount of Br in specimen G was known, it was also possible to compute the recovery of Br for this specimen, which was 71%.

3.1.5.2 Business Machine Housings

For the business machine housings, the FR specimen A showed a slightly longer time to ignition and a slightly lower percent of mass burned. The peak rate of heat release was reduced by some 40% from the un-retarded specimen F. The FR specimen also showed total heat released of some 1/3 lower, with a 30% decrease in the effective heat of combustion. In other words, the mass loss rates for the two specimens were similar, with the lower rate of heat release (RHR) of specimen A being accounted mainly by the lowered heat of combustion (Δh_c). The CO yield rate was about 50% higher for the FR specimen, while the smoke values were essentially identical. There were no qualitative differences in any measured variables between the 30 and the 100 kW/m² test conditions.

3.1.5.3 Upholstered Chairs

For the fully-composite specimens of the upholstered chairs, the FR specimen S took more than twice as long to ignite as the NFR specimen T, and lost a smaller fraction of its mass. The peak rate of heat release was decreased by some 40%, while the total heat released was unchanged and the heat of combustion decreased by about 35%. CO and HCN yields were more than doubled for the FR specimen, compared to the NFR one, while the smoke yield was essentially unchanged. CO_2 yields for the NFR specimens were about half of those for the FR ones. There were no significant differences between the results at the two test irradiances used.

When tested without the cover fabric, the heat and mass loss values were roughly similar to those for the composite specimens, with the advantages of the FR material seen a bit more clearly in some cases. The ignition time for the FR material was 12.5 times longer than for the NFR material. Smoke production, however, was substantially lower for both T and S under the conditions of no cover fabric, suggesting that the fabric's performance here will dominate the results for the composite. HCN yield was similar for the uncovered T as for the covered one, but the FR version, specimen S, did not show the sizeable increase over the NFR one, as was seen in the composites testing. As already reported [11], this again suggests that small, detailed changes of polyurethane fire exposure can make significant changes in the yields of HCN from polyurethane foams.

Since the amount of Cl in specimen S was known, it was possible to compute the recovery of Cl, which was 84%.

For the subsequent large-scale fire tests, it was also necessary to determine the minimum irradiance necessary to cause ignition of the FR polyurethane foam/nylon fabric assembly. These data were obtained in the Cone Calorimeter and are shown in Table 5. The results show that 11 kW/m² is the approximate threshold of ignition for sample S.

3.1.5.4 Cable Arrays

For the cable arrays, when the fully composite specimens were tested, little difference was seen between the specimens when tested at 30 kW/m² irradiance. At 100 kW/m² irradiance, however, the FR specimen showed the same peak rate of heat release as it did at 30 kW/m² irradiance (which is an unusual occurrence). The NFR specimen D behaved more typically in that its rate of heat release at the higher flux was substantially higher than at the lower flux.

Since the amounts of Br and Cl in specimen D were known, recovery fractions could be computed; these were 27% for Br and 105% for Cl. The Cl result can be considered to be approximately 100%, to within experimental uncertainty. The exact halogen content for specimen K was not known.

When the wire specimens alone were tested, however, the FR specimen K showed only about 40% of the heat release rate of the NFR specimen D. Ignition times and mass burned remained essentially identical, while the total heat released and the heat of combustion dropped to half. Smoke yields increased by some 40%, while the CO yield increased by some 4-1/2 times. The results at 100 kW/m² irradiance were not substantially different from the values at the lower heating flux.

The jacketing materials, when tested alone, did not show differences between D and K. This was as expected, since the formulation of the jacketing material was identical for both specimens.

3.1.5.5 Circuit Boards

For the circuit board materials, the FR specimen L showed a peak rate of heat release that was less than that for the NFR specimen C by a factor of 2.5. The ignition time showed a 50% improvement, while the total heat release and the effective Δh_c were also both decreased. Smoke production for the FR specimen was slightly less, while the CO production was substantially higher. At the higher, 100 kW/m² heating flux, the main substantive difference was a near-doubling of the smoke production propensity shown by the NFR specimen; by contrast, irradiance changes did not affect smoke production for the FR specimen.

3.2 FURNITURE CALORIMETER

With the above bench-scale data in hand, it is tempting to make extrapolations to large-scale behavior. The tools for doing so, for most products, are as yet unavailable. Thus, it was necessary to also study the same test articles in a large-scale configuration. These large-scale data can then, validly, be used as input into predicting the behavior of a room fire. In those few cases where methods for predicting large-scale behavior based on bench-scale data are available, direct comparisons between large-scale predictions, based on the bench-scale data and the actual, measured large-scale Furniture Calorimeter data can be compared.

3.2.1 EXPERIMENTAL TECHNIQUE—GENERAL

The NBS Furniture Calorimeter has been developed as a technique for easily measuring the rates of heat release of free-standing combustibles, burning in open-air conditions [2]. As with the Cone Calorimeter, the design of the apparatus is based on using oxygen consumption as the measuring principle. Since it is intended to measure the burning of large objects which can sustain their own burning, once suitably ignited, no external radiant heat is provided. (By contrast, in the Cone Calorimeter, where small-scale specimens are measured, external radiation is necessary in order to represent the effects on the specimen of adjacent items or portions of the item burning.)

Similar to the Cone Calorimeter, the Furniture Calorimeter also contains a load cell, a laser photometer for measuring smoke, and gas analysis equipment. The Furniture Calorimeter currently exists at NBS in two versions, a low-capacity version (for fires not much over 1000 kW) and a high-capacity version (for up to about 7000 kW). The low-capacity version was used in this study. The ignition source, except as specified otherwise, is typically a natural gas burner, having a nominal 180 × 150 mm face and operated at the 50 kW level for 200 s. Those conditions approximate the fire behavior of a small, trash-filled plastic wastebasket [12].

Demonstration tests have shown that under certain conditions the open-air large-scale measurements made in the Furniture Calorimeter can be directly transferred to the room-fire situation [13]. The limits of this relationship have not been fully tested, however. In the present program one of the objectives was to be able to compare the success of this type of prediction.

A summary of the test specimens evaluated with the NBS Furniture Calorimeter and their test conditions are given in Table 6. A description of the test set-up and procedure used for testing each item is given below. Schematics of the test arrangements are given in Figures 9-13. The test results are summarized in Table 7.

3.2.2 EXPERIMENTAL TECHNIQUE—SPECIFIC DETAILS

With the exception of the cable tests 19-21, each of the test specimens was burned under the Furniture Calorimeter using the standard burner described above. Cable tests 19-21 used a 0.36 m long line burner to administer the same flow of natural gas.

Each of the simulated business machine and television cabinets was made by assembling one from two molded half-sections, forming a single closed cabinet with an opening at one end. The overall size of the cabinet was 0.36 × 0.33 × 0.25 m high, and the opening was closed with a galvanized steel cover. No internal working parts were used in the cabinets. In each test, two cabinets, separated with a 25 mm spacing, were placed over the test platform with the burner positioned 0.30 m directly under the spacing between cabinets. Figure 9 shows the test arrangement. The burner was left on for the first 200 s, and the tests were run for 600 s.

Each of the chairs was fabricated with a metal frame and four upholstered cushions (0.61 × 0.61 × 0.1 m). Thus, the chairs comprised only two combustible components, the cushion padding and the previously described upholstery fabric. They were tested with the gas burner positioned along one side of the chair and 0.30 m above the test platform floor. Figure 10 shows the test arrangement. The burner was left on for 200 s, and the fire was allowed to continue until all flames extinguished on the chair.

In cable tests No. 5 and 6, the cables were mounted along a "Z"-shaped 0.46 m wide steel ladder. The upper and lower horizontal sections were 0.53 m long each and the vertical section was 1.37 m long. Thirty five cables were strung side by side across the width on one side of the ladder with a half cable diameter spacing away from each of the two edges of the ladder. The cables filled the entire 2.44 m length of the ladder. The burner was placed 0.30 m directly under the bottom of the vertical section. Figure 11 shows the test arrangement. The burner was left on for 200 s initially and was reignited whenever self-extinguishment occurred. The duration of each re-ignition period was chosen such that there was subsequent vigorous fire involvement of the cables before the burner was turned off again.

Subsequent tests 19-21 of the cables were conducted using a vertical ladder array of 25 cables, 2.18 m high and placed side by side for a total width of 0.30 m. Test 21 was only a test of the cable jackets. The inner wires were removed and replaced with a solid copper wire with a diameter of 0.08 m. The test arrays were exposed for 1200 s to a 0.36 m wide line burner, which was chosen to provide greater flame coverage of the cable array surface. The line burner was positioned 60 mm below the bottom of the cables with an offset of about 13 mm towards the front of the cable array such that the burner flames would initially contact the front surface of the cables. Figure 12 shows the test arrangement. The fire was allowed to continue until all flames extinguished on the cables.

In the circuit board tests, eight 0.91 m by 0.41 m by 6.4 mm thick specimen panels were placed in a vertical array with 25 mm spacing between panels. The gas burner was located 0.30 m directly below the array. The cable array and the burner were housed in a 0.51 × 0.51 × 1.91 m high metal relay-rack type cabinet, which had the top and lower 0.61 m of the front open for ventilation. Figure 13 indicates the test arrangement. The burner was left on for the first 200 s and the test specimen panels were allowed to burn up to 2500 s unless the fire extinguished sooner. For materials which self-extinguished during the test, the burner was re-ignited for various periods of time to further assess the materials under longer fire exposure times.

3.2.3 GAS SAMPLING THROUGH IMPINGERS

Samples were collected from the exhaust gas stack on a platform above the Furniture Calorimeter hood. A diagram of the gas sampling apparatus is shown in Figure 14. A portion of the stack gases was drawn through the impinger assembly with a sampling pump controlled at nominally 1000 ml/min with an electronic mass flow controller. The flow rates through the impingers were verified with a laboratory dry-gas meter. The ratio of gases collected to gases exhausted was approximately 1:100,000. Collection of sample began when the specimen on the load platform was ignited; collection was stopped when the flaming stopped and no appreciable smoke was seen.

The same PTFE filters and gas sampling impingers previously mentioned were used. The solution in the impingers was 5 mM KOH. The filters containing soot were placed into the solution from the first impinger. The samples were then prepared and analyzed as described above.

3.2.4 FURNITURE CALORIMETER RESULTS

The Furniture Calorimeter results are summarized in Table 7. The RHR curves for the tests are given in Figures 15 to 19. In each case, the FR product was notably better-behaved in its *peak* rate of heat release characteristics. In some cases (the business machine and TV housings, the circuit boards, and the cable arrays in the Z-arrangement) the FR product showed, very roughly, half the RHR of the un-retarded one. In other cases (the chairs and the cable arrays in the vertical arrangement) the FR product showed no continued flame propagation at all, leading to RHR values which are much less, but not strictly comparable to, those registered by the NFR product. Since it might take just a little more external heating in order to show complete and continued flame spread, the Furniture Calorimeter test is not necessarily the final and definitive test for those FR products which fail to show continued flame spread when tested there. Instead, the need is seen for ultimate validation with an appropriate room fire test. These will be discussed below.

3.2.5 DIRECT COMPARISON TO CONE CALORIMETER RESULTS

Table 8 compares data obtained in the Cone Calorimeter, and those gathered in the Furniture Calorimeter. The simplest comparison is between the bench-scale and the large-scale rates of heat release. The quantities measured are, of course, not reported in the same units, since in bench scale the value determined is the heat release per unit area (kW/m^2), whereas in large scale, it is the heat release for the entire object (kW). Thus, only the relative magnitudes must be compared, not the absolute numbers. For the business machine housings, the television cabinets, and the circuit boards, the comparison is consistent to within about 25%.

For the chairs and for the cables a good comparison is not obtained. The analysis of the chair performance is given in the following section. For the cables, the bench-scale tests run on the composite specimens give almost identical results for the FR product K and the NFR product D. This is not surprising, since when complete, composite specimens are tested, the behavior is dominated by the outer rubber jacket. This jacket is the same for both specimens. In the Furniture Calorimeter, however, the FR specimen K showed a substantially better behavior than the NFR specimen D. This may be explained the following way. The fire retardant agent in specimen K is only contained in the wire insulation. In the large-scale test rig, the pyrolysis products of the burning section are swept up along the not-yet-ignited upper portions of the specimen. In the case of specimen K, the FR agent is apparently active enough to stop flame spread from proceeding upwards. To examine the plausibility of this conjecture, data from an additional Furniture Calorimeter test were obtained, where the wires had been removed from the outer jacket, and only the jacket remained as a combustible. (The copper tube placed inside the jacket served to preserve its dimensional stability.) These data are shown in Table 7. The peak $\dot{q}$ for specimen K was 75 kW, while the peak $\dot{q}$ for the jacket-only test was 140 kW. Thus, rate of heat release for the jacket alone is greater than for the actual assembly of jacket + wire, supporting the hypothesis that the pyrolyzed FR agent from the wire insulation actually stops the flame spread and burning of the jacketing material.

For other variables of interest, in those cases where the large-scale specimens did show continued flame spread and combustion in the Furniture Calorimeter, the fraction of the combustible mass lost was similar to that observed in the bench-scale (Table 8). For specimens, such as the FR circuit board or the FR chair, where continued flame spread over the specimen surface was not sustained, the fraction of mass lost in the Furniture Calorimeter was very small and was not comparable to the fraction lost in the Cone Calorimeter. The circuit board L presents an interesting case; the total mass lost and the total heat recovered were very small, yet a peak value of around 100 kW rate of heat release was observed, which is not insignificant.

Comparing gas yields, in most cases the yield of CO was higher in the Furniture Calorimeter than in the Cone Calorimeter. The FR products typically showed higher CO yields in the Furniture Calorimeter, compared to the NFR controls, but the relative NFR vs FR differences in the Furniture Calorimeter were much smaller in most cases than in the Cone Calorimeter. It is interesting to note that the range of fractional CO yields per unit mass of fuel burned in the Furniture Calorimeter was quite small, with most products, both NFR and FR, being clustered at around 0.10 to 0.13.

The yields of CO_2 in the Furniture Calorimeter were closely comparable to those in the Cone Calorimeter, with the exception of specimens H, F, and L. Since two of these are NFR products, and one is FR, no significant difference between the predictability of CO_2 yields of NFR vs FR products is seen.

The acid gas (HCl, HBr, HCN) measurements show very close agreement in all cases where data from both instruments were available. Specimen T, while disagreeing by 50% between the tests, in fact shows small HCN yields, comparable to the measurement limit. Data on recovery fractions of halogens were available in both the Cone Calorimeter and in the Furniture Calorimeter only for Br for specimen G. In that case, both sets of results indicated a 71% recovery.

For smoke yields, the FR products behaved either very similar to the NFR ones (business machine housings, cables, upholstered chairs), or better (circuit board). The one anomalous case is of the TV cabinet housing where smoke yields were about double for the FR specimen G, as compared to the NFR specimen H. Since bench-scale tests also showed this 2:1 difference, the finding seems validated.

The relationship of NFR to FR smoke yields was very similar in both the Cone Calorimeter and the Furniture Calorimeter in most all cases. Only in the case of the FR circuit board specimen L was this prediction poorly held. This is very simply attributed to the fact that the amount of specimen L which burned in the Furniture Calorimeter was very small, therefore, yields of all products are highly uncertain. Comparing now the numerical values of the average smoke extinction area obtained in the Cone Calorimeter and in the Furniture Calorimeter, the values were typically less in the Furniture Calorimeter by a factor of 2 to 3. This scale effect has been already noted in connection with earlier studies on upholstered furniture [8]. In the simplest case, one might suppose that the large-scale and the bench-scale values ought to be identical.

Some recent data, in fact, suggest that identity between bench-scale and large-scale smoke values can be achieved if the $\dot{m}''$ (mass loss rate, per unit area) values of the specimens are matched to be identical for both cases [14]. Such matching can generally be done by varying the external irradiance in the bench-scale test. In that study [14], however, the articles were primarily burning in a fully-flame-involved manner for nearly the entire duration of test. In the present study (and also in the earlier upholstered furniture study), the period of flame spread over the surface tended to comprise a large fraction of the total burning time. Thus, it is possible that the reason why neither of these studies shows such 1:1 matching might be due to flame spread effects. The one exception to the above was for the TV cabinet housings, where both the FR specimen and the NFR one showed a greater smoke yield in the Furniture Calorimeter than in the Cone Calorimeter.

3.2.6 USE OF PREDICTIVE SMALL-SCALE/LARGE-SCALE CORRELATIONS

As stated above, in general correlations have not yet been developed for most classes of products which would permit the engineer to accurately estimate the likely large-scale fire performance, based on bench-scale data alone. In some cases, however, such inroads have started to be made. A recent review [15] summarizes the state of the art for such predictions for the rate of heat release ($\dot{q}$) and the mass loss rate ($\dot{m}$). Of the products included in the present test series, a predictive correlation is available for upholstered chairs. This correlation was derived for residential type of furniture constructions, whereas the test program involves materials used for office occupancies. Nonetheless, it becomes interesting to examine the possible applicability of this correlation. According to this prediction [15], the peak rate of heat release is estimated to be

$$\dot{q} = 0.63 \, \dot{q}''_{bs} \, [\text{mass factor}][\text{frame factor}][\text{style factor}] \quad (1)$$

The $\dot{q}''_{bs}$ is the bench-scale rate of heat release (kW/m²), averaged over a 180 s interval, starting at the time of ignition, on the composite foam/fabric specimens. For this correlation, the prescribed irradiance is 25 kW/m². Although 25 kW/m² irradiance tests were not conducted under the present program, only a very small error is expected if the data from the 30 kW/m² irradiance tests are used, instead. The mass factor is equal to the total combustible mass of the large-scale article (kg). For the steel frames used in the present tests, the frame factor is 1.66. The style factor is 1.0 for the rectilinear construction used.

The measured 180 s average $\dot{q}''_{bs}$ values at 30 kW/m² irradiance were:

310 kW/m² for specimen T
165 kW/m² for specimen S

Thus, for the NFR chair specimen T, the predicted peak value of large-scale rate of heat release is $0.63(310)(5.5)(1.66)(1.0) = 1800$ kW. For the FR chair specimen S, meanwhile, the predicted value is $0.63(165)(11.9)(1.66)(1.0) = 2100$ kW. By contrast, the measured values were 1200 kW for chair T and 50 kW for chair S. Thus, the agreement is not very good.

The lack of agreement deserves some analysis. The tests in the present test series were run at an irradiance of 30 kW/m², instead of the 25 kW/m² used for the original correlation. The numerical effect of this is expected to be small, in the range of 5 to 10%, yet specimen T is over-predicted by some 50%.

For specimen S, the concern is of a very different sort. The predictive correlation above is based on the premise that flame spread covers all of the specimen, and that essentially the whole combustible mass is consumed. In fact, sustained flame spread was not observed over specimen S, and only a negligibly small fraction of it was consumed. Some earlier work has already identified that there is a limiting value of $\dot{q}_{bs}''$, below which continued flame spread is not expected, and, therefore, Equation 1 is not expected to be applicable [16]. This limiting value has not been well-characterized; a value of around 75 kW/m² was suggested in [16]. Quintiere suggests a value of around 100 kW/m² [17]. Specimen S, however, showed no sustained flame spread while having a $\dot{q}_{bs}''$ value of 165 kW/m². This suggests either that (a) a significantly higher no-spread criterion value may have to be considered; or, (b) that the no-spread condition was due to very localized burning peculiarities adjacent to the igniting burner, and that a different burner arrangement might, in fact, have led to sustained spread. Without doing additional testing, the proper cause could not be established.

3.3 SMALL-SCALE TOXICITY TESTS

The objective of this series of tests was to determine whether the results of the small-scale tests would predict the toxicity (based on lethality) of the large-scale tests. The smoke toxicity in both the small-scale tests and large-scale tests is predicted by the current N-Gas Model which is based on data from earlier NBS experiments on the combined toxicity of the major gases produced in fires [3,18]. The N-Gas model is designed to determine whether the lethality of the thermal decomposition products of a material can be explained by the toxicological interaction of these major gases or if other toxic combustion products need to be considered. The N-Gas model also permits the prediction of approximate LC_{50} values based on the measured concentrations of the predetermined set of gases and knowledge of their toxicological interactions. The LC_{50} value is defined as the concentration which is lethal to 50% of the test animals for a specific exposure time. Animal exposure experiments are conducted at the predicted LC_{50} value and if the prediction is correct, some percentage (not including zero or 100%) of the animals will die and the approximate LC_{50} value is assumed to be close to the actual LC_{50}. Death of all the animals indicates that other toxic gases are contributing to the toxic atmospheres. Currently, four gases, CO, CO_2, HCN, and low O_2, have been examined both individually and in various mixtures and are included in the model. (For some estimates below, however, approximate expressions for HCl and HBr effects are included, based on data reported by others.)

The bioassay based on the N-Gas model, thus, provides information on the toxic potency of the combustion products from a material which can be used to answer two questions:

(1) Does the material show normal toxic potency? That is, does it show a value of the LC_{50} which is comparable to products ordinarily found in a built environment? For this study we shall consider any substances showing $LC_{50} > 2$ mg/l as having normal toxic potency. Conversely, any product showing $LC_{50} < 2$ mg/l would be considered as having extreme toxic potency.

(2) Does the material show unusual toxicity? That is, is any significant fraction of the toxicity not accounted for by the gases being monitored (CO, CO_2, HCN, low O_2, HCl, and HBr).

In this phase of the work, toxicity experiments involving both bioassay and analytical gas measurements were conducted to test the predictive capabilities of the N-Gas Model. The materials were examined under flaming conditions using the NBS toxicity test apparatus (Figure 20) [19].

3.3.1 ANIMALS

Fischer 344 male rats, weighing 200–300 grams, were obtained from Taconic Farms (Germantown, NY). They were allowed to acclimate to our laboratory conditions for at least 10 days prior to experimentation. Animal care and maintenance were performed in accordance with the procedures outlined in the National Institutes of Health's Guide for the Care and Use of Laboratory Animals. Rats were housed individually in suspended stainless steel cages and provided with food and water ad libitum. Twelve hours of fluorescent lighting per day were provided using an automatic timer. All animals (including the controls) were weighed daily from the day of arrival until the end of the post-exposure observation period.

3.3.2 SMALL-SCALE TOXICITY TEST SYSTEM

The acute inhalation toxicity of the combustion products of the above materials was assessed using the N-Gas Model Toxicity Procedure (Table 9) which uses the combustion system, the chemical analysis system and the animal exposure system that was designed for the NBS Toxicity Test Method [19]. This test method provides for both flaming and non-flaming exposures. Since the large-scale test scenarios envisaged for the present program did not involve any non-flaming conditions, only flaming exposures were conducted. Analytical experiments without animals were first conducted in order to determine the autoignition temperature and the concentrations of CO, CO_2, and HCN that are generated from different mass loadings of each material. These gas concentrations were then used to predict an LC_{50} value based on the N-Gas model. Where possible, the animals were then exposed to a mass of material which was equal to this predicted value. In certain cases, however, the animals could only be exposed to the highest concentration permitted by the physical limitations of the combustion system.

The animal exposure system is a closed design in which all the gases and smoke are kept in the 200 l rectangular chamber for the duration of the experiment. The materials are decomposed in a cup furnace located directly below the animal exposure chamber such that all the combustion products from the test materials are evolved directly into the chamber. For this study, the materials were tested under flaming conditions, achieved by setting the furnace 25°C above each material's predetermined autoignition temperature. When needed, ethanol or an electric spark was used to ensure immediate flaming.

In the animal exposure experiments, six rats were exposed in each experiment. Animals were placed in restrainers which were then inserted into the six portholes located along the front of the exposure chamber such that only the heads of the animals were exposed. The animals were exposed to the test atmospheres for 30 minutes. This exposure comprised an initial period (which usually was about 5 minutes) during which the test specimen was actively decomposing, as well as the later steady-state condition. The toxicological endpoint was death which occurred either during the 30 minute exposures or the post-exposure observation period. This period was usually 14 days, but if animals were losing weight, they were kept until they either died or recovered, as indicated by three days of weight gain.

3.3.3 GAS ANALYSIS

Carbon monoxide and carbon dioxide were measured continuously by non-dispersive infrared analyzers. Oxygen concentrations were measured continuously with a paramagnetic analyzer. The CO, CO_2, and O_2 data were recorded by an on-line computer every 15 seconds. All combustion products and gases that were removed for these analytical measurements were returned to the chamber. The HCN generated from the nitrogen-containing polymers was sampled approximately every three minutes with a gas-tight syringe and analyzed with a gas chromatograph equipped with a thermionic detector [20]. These gas concentrations are provided as the time-weighted average values which were calculated for the exposure period.

The presence of HBr in the gaseous combustion products was determined by bubbling a portion of the gases generated in the animal box through 3 ml impingers. The apparatus is diagrammed in Figure 21. The PTFE filter (0.45 μm nominal porosity) in this test was 25 mm in diameter. The flow was generated with a sampling pump and regulated with a rotameter. The flow was nominally 30 ml/min and verified prior to each test with a soap bubble meter attached to the tube that protruded into the box. The duration of the test was 30 minutes. In this test, the impingers contained approximately 2 ml of 5 mM KOH. After the test, the PTFE filter containing the soot was placed in the first impinger and the contents of each impinger were brought to 5 ml. The samples were then analyzed for bromide ion as previously described.

3.3.4 N-GAS MODEL PREDICTION

The current N-Gas model is based on studies of the toxicological interactions of four gases [3,18]. The following equation has been experimentally derived to predict when 50% of the animals should die either within the 30 minute exposures or within a 24 hour post-exposure period. The current model does not predict post-exposure deaths occurring after 24 hours.

For CO, HCN, CO_2 and low O_2:

$$\frac{m[CO]}{[CO_2] - b} + \frac{[HCN]}{LC_{50} \; HCN} + \frac{21 - [O_2]}{21 - 5.4} \approx 1 \quad (2)$$

where brackets indicate the actual concentration of the gases; the LC_{50} value of CO is based on 30 min exposures and is 6600 ppm; the LC_{50} value of HCN is 160 ppm for deaths occurring within the 30 min exposure or 110 ppm for deaths occurring during the exposure or in the subsequent 24 hour post-exposure observation period; and 5.4 is the percent O_2 that causes 50% of the animals to die in 30 minutes. The terms m and b equal -18 and 122,000 if the CO_2 concentrations are 5% or less and 23 and $-39,000$, respectively, if the CO_2 concentrations are above 5%. The period under consideration here is restricted to 24 h, since delayed deaths due to HCN are usually not found after the first 24 h.

Since the concentration-response curves for animal deaths from combustion products are very steep, the assumption is made that if any percentage of the animals die (not including 0 or 100%), the concentration should be close to the LC_{50} value. Examination of a series of pure gas experiments in which various percentages of the animals died indicated that the mean N-Gas value was 1.07 with a standard deviation of 0.10. This means that the 95% probability that some animals will die as a result of exposure to these primary gases falls within the range of 0.87 to 1.27. Deaths below this range may then be attributed to the additional toxicity contributed by other gases or factors. Above this range, all the animals would be expected to die.

3.3.5 RESULTS AND DISCUSSION OF THE SMALL-SCALE TOXICITY TESTS

All the bench-scale chemical and toxicological data obtained for the test specimens in the N-Gas Model Toxicity Procedure are presented in Table 10. With no animals present, each material was first tested in the flaming mode (25°C above its predetermined autoignition temperature) at a mass loading of 40 mg/l to obtain the necessary analytical gas data to calculate the value of the 4-Gas Model prediction. In Table 10, the analytical data for each material are given first, before the experiments in which animals were exposed. For ease in interpretation of the present test results, toxicity results for a number of common materials previously studied [19] are summarized in Figure 22.

When 40 mg/l (mass loading or mass consumed-chamber volume) of the NFR television cabinet (Specimen H) was decomposed in the flaming mode, the 4-Gas Model prediction value ranged from 0.93 to 0.99 and one of the six exposed animals died during the exposure. When the mass loading and mass consumed was reduced to 30 mg/l, no deaths were noted. Therefore, the estimated LC_{50} value is approximately 40 mg/l and the toxicity of the combustion products of this material would not be considered extremely toxic. Since the 4-Gas Model prediction value is approximately 1.0 at the LC_{50}, the toxicity of the measured gases alone is considered responsible for the observed deaths.

Similar to its NFR counterpart, the FR television cabinet (Specimen G) decomposed in the flaming mode was completely consumed during the 30 minute exposure. Therefore, the mass loading/chamber volume is equivalent to the mass consumed/chamber volume. However, this FR specimen did not maintain the flaming condition unless the ignition spark was left on continuously. Consequently, with the exception of the first experiment, all the tests on this material were conducted with the ignition spark on throughout the exposure. At 40 mg/l, the 4-Gas Model prediction value was higher (1.42–1.54) in the cases where the sparker was left on compared to the test (0.65) in which the sparker was turned off after the material ignited. At this 4-Gas Model prediction value, all the animals died as expected. At 20 mg/l, the 4-Gas Model prediction value was 0.56 and no animals died either during or following the exposures. At 30 mg/l, the 4-Gas Model prediction value was 0.79 and while no animals died during the exposure, two died within the first 24 hours of the post-exposure period. These results indicate that the estimated LC_{50} value is close to 30 mg/l; therefore, the toxicity of the combustion products of this material is not extremely potent. However, since the 4-Gas Model prediction value at the estimated LC_{50} is 0.79, which is below the lower 95% confidence limits of the range where deaths are expected, another gas or gases may be contributing to the post-exposure deaths. The average 30-minute concentration of HBr in this experiment was 163 ppm. At the present time, HBr is not one of the gases included in the 4-Gas Model; however, if one considers the toxicological contribution of this gas as an additional additive factor and assumes the 30-minute LC_{50} value as 3000 ppm [5], another 0.05 would be added to the 4-Gas Model prediction value (i.e., 0.84). This value of the 4-Gas Model is now fairly close to the lower 95% confidence limits and it is uncertain as to whether a further gas is contributing to the toxicity of these combustion atmospheres.

At a 40 mg/l loading, the value of the 4-Gas Model prediction for the NFR business machine (Specimen F) was determined as 1.47. Consequently, the loading of material into the furnace was decreased such that the value of the 4-Gas Model prediction was 1.08, a value at which deaths of some of the animals would be expected. In actuality, no animals died either during or post-exposure from this experiment. Therefore, the mass loading was increased such that the 4-Gas Model prediction value was 1.36; at this loading, all of the animals died within the exposure as expected. These data indicate that the LC_{50} value is between 32 and 35 mg/l mass loaded (30 to 33 mg/l mass consumed) or in terms of the 4-Gas model prediction values, between 1.08 and 1.36. Therefore, the toxicity can be attributed to the series of gases that were monitored. Thus, this material is neither extremely toxic nor unusually toxic (i.e., gases other than the major ones measured are not contributing to the toxicity).

The chemical and animal exposure results from the FR business machine housing (Specimen A) indicated that the estimated LC_{50} value was about 25 mg/l based on mass loaded (24 mg/l based on mass consumed) and that the FR specimen is slightly—but not toxicologically significantly—more toxic than the NFR specimen. Since 4 out of 6 animals died at an exposure in which the 4-Gas Model prediction value was 1.13, this material is also not unusually toxic.

The NFR flexible polyurethane foam corresponding to Specimen T, when decomposed in the flaming mode at a mass loading/chamber volume of 40 mg/l (35 mg/l mass consumed/chamber volume), produced 4-Gas Model prediction values of 0.55 or 0.60 for within- and post-exposure deaths, respectively (the difference between these two numbers depends upon the value of the LC_{50} used for HCN). No deaths occurred from this exposure indicating that the LC_{50} value of this material is >35 or >40 mg/l depending on whether the mass consumed or mass loaded-chamber value is used for the calculation. Therefore, the toxicity of the combustion products would not be considered extremely potent. Since experiments could not be conducted at loadings closer to the LC_{50} value, it was not determined if the toxicity of lethal levels of the combustion products can be explained by the combined measured gases or if other gases are needed. However, the contribution of additional gases would only be <30%.

The FR flexible polyurethane foam, corresponding to Specimen S, did not burn in a consistent manner; hence, there were a number (4) of analytical runs (without animals) at a mass loaded/chamber volume of 40 mg/l (24-26 mg/l mass consumed/chamber volume). Since all but one of these tests were at 550°C, only the 550°C tests will be discussed. Even at the same mass loading (40 mg/l), the 4-Gas Model prediction value varied between 0.74 and 1.10 (calculated for within-exposure deaths) or 0.80 and 1.23 (calculated for post-exposure deaths). Two animal experiments were performed at mass loadings/chamber volume of 30 and 35 mg/l (20 and 23 mg/l mass consumed/chamber volume, respectively). No animals died during either of the exposures; however, two animals died during the first 24 hours following exposure to the combustion products from the experiment with the higher mass loading. The estimated LC_{50} value is approximately 35 mg/l or 23 mg/l based on mass loaded or consumed/chamber volume, respectively. The toxicity of the combustion products would not be considered extremely potent. Since only two animals died post-exposure at a 4-Gas Model Prediction value of 1.23, the toxicity of these combustion products can be attributed solely to those gases that were monitored.

The NFR wire insulation, corresponding to Specimen D, decomposed in the flaming mode at a mass loading/chamber volume of 40 mg/l (34 mg/l mass consumed/chamber volume) produced a 4-Gas Model Prediction value of 0.66. Unless other gases were contributing to the toxicity, this value would not be expected to cause any deaths and, in fact, no deaths were noted in the corresponding experiment. When the amount of material loaded (and consumed) was doubled, the 4-Gas Model prediction value was 1.47 and as expected, all of the animals died during the exposure. The estimated LC_{50} is between 40 and 80 mg/l based on mass loaded/chamber volume or between 34 and 68 mg/l based on mass consumed/chamber volume. These LC_{50} values indicate that the toxicity of this material is not extremely potent. Due to the screening nature of this experimental series, it is unclear whether additional gaseous agents need to be sought.

When 40 mg/l (mass loaded/chamber volume or 32 mg/l mass consumed/chamber volume) of the FR wire insulation (from Specimen K) was decomposed in the flaming mode, the 4-Gas Model prediction value was 0.71 when no animals were present and 0.81 when animals were exposed (the animals can contribute to the CO_2 concentrations and reduce the O_2). At this loading, one animal died during the exposure and all the animals died within 24 hours following the exposure. Although half of this loading produced a 4-Gas Model prediction value of only 0.41, one animal died during the post-exposure observation period on day 19 (Figure 23). Since the LC_{50} of this material is between 20 and 40 mg/l based on mass loaded/chamber volume or between 17 and 33 mg/l based on mass consumed/chamber volume, the toxicity of the combustion products would not be considered extremely toxic. However, since the 4-Gas Model prediction value at the LC_{50} is between 0.41 and 0.81, one or more gases, other than those monitored, are contributing to the toxic combustion atmospheres. Procedures for including HCl toxicity in the NBS Toxicity Test Method have not yet been worked out. Since the main additional toxic agent expected would be HCl, it is possible to estimate its expected effect. The yield of HCl in the Cone Calorimeter for this specimen was 0.095 kg/kg (Table 2). If this same HCl yield is also realized in the NBS Toxicity Test Method, then HCl concentration within the box would be 2500 ppm. Using 3700 ppm as the 30-minute (+14 day post-exposure) LC_{50} value for HCl [5], the contribution of HCl would be 2500/3700 = 0.68. If linear additivity holds for HCl in conjunction with the other gases, this would change the prediction value from being 0.41 to 0.81 to being 1.09 to 1.49. This range is within the area where animal deaths (primarily occurring post-exposure, due to HCl) are successfully accounted for. Additional research, however, would be needed to confirm this prediction.

The NFR circuit board (Specimen C), when decomposed at a mass loading/chamber volume of 41 mg/l (12 mg/l mass consumed), produced very low amounts of the monitored gases. The 4-Gas Model prediction values for the experiments without and with animals were only 0.18 and 0.21, respectively. This indicated that only about 20% of the toxic gases had been produced by this loading of material. An increase in the loading such that the 4-Gas Model prediction value was approximately unity (1.08) resulted in no deaths during or following the exposure. However, an increase in the loading that produced a 4-Gas Model prediction value of 1.38 caused all the animals to die within the exposure period. From these data, a LC_{50} value based on mass loaded/chamber volume can be estimated as between 165 and 200 mg/l; if the LC_{50} value is based on mass consumed/chamber volume, then the value is between 48 and 58 mg/l. Either way the toxicity of this material would not be considered extremely potent. Since the 4-Gas Model prediction value at the LC_{50} value is between 1.08 and 1.38, this material is not unusually toxic.

The FR circuit board (Specimen L) when decomposed in the flaming mode at a mass loading/chamber volume of 40 mg/l (19 mg/l, mass consumed/chamber volume) produced concentrations of CO, CO_2, and O_2 which resulted in a 4-Gas Model prediction value of approximately 0.40. This 4-Gas value would not be expected to produce deaths and no deaths occurred. When this mass loading/chamber volume was doubled (85 mg/l loaded, 36 mg/l consumed), the 4-Gas Model prediction value increased to 0.76, which is still below the amount necessary to cause deaths if those deaths are due to the gases being monitored. Again, no deaths occurred. At a mass loading of 100 mg/l (45 mg/l consumed), the 4-Gas Model prediction value was 1.13 and all the animals died. This indicates that the estimated LC_{50} value lies between 85 and 100 mg/l based on mass loading or between 36 and 45 mg/l based on mass consumed. In either case, the toxicity of the flaming combustion products of this material would not be considered extremely potent. Since all the animals died at a 4-Gas Model prediction value of 1.13 and none died at 0.76, it is unclear what the 4-gas prediction value would be at the LC_{50}, and, thus, it is not known whether other gases are contributing to the toxicity in this case.

The results of the small-scale toxicity tests are summarized in Table 11.

TABLE 1 Cone Calorimeter Data Summary—30 kW/m² Irradiance Tests.*

Sample	NFR/FR	Mass (g)	% Mass burned	Ign. Time (s)	Peak $\dot{q}''$ (kW/m²)	Peak time (s)	Tot. q'' (MJ/m²)	Eff. Δh_c (MJ/kg)
TV Cabinet H	NFR	34	99	107	970	190	87	30
TV Cabinet G	FR	38	98	84	340	184	46	12
Bus. Machine F	NFR	37	88	108	650	168	96	30
Bus. Machine A	FR	39	81	134	380	370	65	21
Chair T[a]	NFR	23	89	14	470	113	54	27
Chair S[a]	FR	43	67	34	290	51	51	18
Chair T[b]	NFR	15	90	2	540	65	34	27
Chair S[b]	FR	36	61	25	180	—	32	15
Cable D	NFR	166	35	383	360	505	156	28
Cable K	FR	170	33	374	380	487	114	23
Cable D[c]	NFR	54	52	189	270	208	65	23
Cable K[c]	FR	53	54	169	280	185	68	23
Cable D[d]	NFR	103	22	137	740	280	91	39
Cable K[d]	FR	106	22	131	260	161	51	23
Circuit Bd. C	NFR	123	28	199	250	220	73	21
Circuit Bd. L	FR	117	36	315	100	368	55	13

*Estimates of uncertainty (median values), expressed as coefficients of variation, are:
  sample mass = 1.9%
  % mass burned = 1.3%
  ignition time = 8.7%
  peak $\dot{q}''$ = 8.6%
  peak time = 6.3%
  total q'' = 2.9%
  eff. Δh_c = 3.1%

[a]Foam and fabric cover combination
[b]Foam only, no cover
[c]Cable jacket only
[d]Wire alone; jacket stripped off

TABLE 2 Cone Calorimeter Data Summary—Test Average Data at 30 kW/m² Irradiance.*

Sample	NFR/FR	CO kg/kg	CO_2 kg/kg	HCl kg/kg	HBr kg/kg	HCN kg/kg	Smoke m²/kg
TV Cabinet H	NFR	0.015	2.284	—	—	—	1010
TV Cabinet G	FR	0.109	0.671	—	0.069	—	1880
Bus. Machine F	NFR	0.037	2.211	—	—	—	1710
Bus. Machine A	FR	0.055	1.604	—	—	—	1660
Chair T[a]	NFR	0.020	1.617	—	—	0.002	410
Chair S[a]	FR	0.051	0.964	0.022	—	0.005	480
Chair T[b]	NFR	0.016	1.711	—	—	0.002	270
Chair S[b]	FR	0.055	0.809	0.022	—	0.002	280
Cable D	NFR	0.041	1.773	0.112	—	—	1010
Cable K	FR	0.060	1.337	0.131	—	—	880
Cable D[c]	NFR	0.029	2.190	—	—	—	690
Cable K[c]	FR	0.135	1.004	0.095	—	—	1030
Cable D[d]	NFR	0.030	2.208	0.128	—	—	710
Cable K[d]	FR	0.142	0.991	0.136	—	—	1000
Circuit Bd. C	NFR	0.014	2.070	—	—	—	560
Circuit Bd. L	FR	0.103	0.868	—	0.022	—	400

*Estimates of uncertainty (median values), expressed as coefficients of variation, are:
  CO = 7.3%
  CO_2 = 7.9%
  smoke = 8.4%
Due to limited amount of data, estimates are not derived for HCl, HBr, and HCN.

[a] foam and fabric cover combination
[b] foam only, no cover
[c] wire alone; jacket stripped off
[d] cable jacket only

TABLE 3 Cone Calorimeter Data Summary—100 kW/m² Irradiance Tests.*

Sample	NFR/FR	Mass (g)	% Mass burned	Ign. Time (s)	Peak $\dot{q}''$ (kW/m²)	Peak time (s)	Tot. q'' (MJ/m²)	Eff. Δh_c (MJ/kg)
TV Cabinet H	NFR	32	97	15	1400	68	93	29
TV Cabinet G	FR	36	95	13	480	55	39	10
Bus. Machine F	NFR	37	88	11	1100	46	95	29
Bus. Machine A	FR	35	87	11	570	41	60	20
Chair T[a]	NFR	22	93	5	1460	52	58	28
Chair S[a]	FR	45	72	5	760	22	56	18
Chair T[bd]	NFR	14	88	<1	1580	35	37	29
Chair S[bd]	FR	37	66	2	310	15	35	14
Cable D	NFR	170	38	8	550	225	159	26
Cable K	FR	173	35	10	380	32	119	21
Cable D[c]	NFR	102	23	16	1280	93	88	38
Cable K[cd]	FR	106	23	16	490	45	50	21
Circuit Bd. C	NFR	127	29	32	250	160	71	18
Circuit Bd. L	FR	116	43	49	147	128	74	14

*Estimates of uncertainty (median values), expressed as coefficients of variation, are:
  specimen mass = 2.7%
  mass burned = 4.1%
  ignition time = 10.7%
  peak $\dot{q}''$ = 6.1%
  peak time = 10.4%
  total q'' = 5.4%
  eff. Δh_c = 3.9%

[a] Foam and fabric cover combination
[b] Foam only, no cover fabric
[c] Wire alone; jacket stripped off
[d] Only one test value

TABLE 4 Cone Calorimeter Data Summary—Test Average Data at 100 kW/m² Irradiance.*

Sample	NFR/FR	CO (kg/kg)	CO_2 (kg/kg)	Smoke (m²/kg)
TV Cabinet H	NFR	0.063	2.121	1430
TV Cabinet G	FR	0.074	0.564	2010
Bus. Machine F	NFR	0.060	1.627	1530
Bus. Machine A	FR	0.096	1.165	2120
Chair T[a]	NFR	0.021	1.828	340
Chair S[a]	FR	0.063	0.965	500
Chair Tb[d]	NFR	0.018	1.889	450
Chair S[bd]	FR	0.052	0.895	420
Cable D	NFR	0.007	1.566	1270
Cable K	FR	0.025	1.245	1210
Cable D[cd]	NFR	0.035	2.148	760
Cable K[cd]	FR	0.101	0.910	1290
Circuit Bd. C	NFR	0.012	1.697	780
Circuit Bd. L	FR	0.012	1.221	410

*Estimates of uncertainty (median values), expressed as coefficients of variation, are:
  CO = 14.3%
  $CO_2$2 = 4.4%
  smoke = 5.6%

[a]Foam and fabric cover combination
[b]Foam only, no cover fabric
[c]Wire alone; jacket stripped off
[d]Only one test value

TABLE 5 Irradiance Threshold Limit for Foam S with Nylon Fabric Cover.

Test Number	Irrad. (kW/m²)	Mass (g)	Mass % burned	Ign. time (s)	Peak q̇″ (kW/m²)	Peak time (s)	Total q″ (MJ/m²)	Eff. Δh^c (MJ/Kg)	CO (kg/kg)	CO_2 (kg/kg)	Smoke (m²/kg)	Soot (kg/kg)
2590	7	43	—	NI	—	—	—	—	—	—	—	—
2592	10	43	14.3	541	180	580	9.6	15.7	0.035	1.558	350	0.056
2593	10	43	—	NI	—	—	—	—	—	—	—	—
2594	11	44	20.5	341	190	370	13.1	14.6	0.035	0.677	380	0.052
2595	11	43	—	NI	—	—	—	—	—	—	—	—
2596	11	42	16.9	527	50	815	5.6	7.8	0.053	0.558	590	0.103
2597	11	43	19.8	264	180	295	13.4	15.7	0.031	0.573	380	0.052
2591	15	43	41.2	173	280	210	22.0	12.5	0.036	0.888	420	0.039

TABLE 6 Summary of Furniture Calorimeter Test Conditions.

Specimen	NFR/FR	Test	Specimen Mass (kg)	Time Period for Burner on (s)	Time Duration (s)
TV Cabinet H	NFR	3	3.7	0-200	600
TV Cabinet G	FR	4	3.7	0-200	600
TV Cabinet G	FR	15	3.7	0-200	660
Bus. Machine F	NFR	1	3.5	0-200	600
Bus. Machine A	FR	2	3.5	0-200	600
Chair T	NFR	16	5.5	0-200	600
Chair T	NFR	18	5.3	0-200	600
Chair S	FR	17	11.9	0-200	345
Cable D[a]	NFR	5	17.5	0-200, 240-600	1400
Cable K[a]	FR	6	18.2	0-200, 240-600, 730-1380	2000
Cable D[b]	NFR	19	11.4	0-1200	1200
Cable K[b]	FR	20	11.5	0-1200	1860
Cable K[c]	FR	21	12.0	0-1200	840
Circuit Bd. C	NFR	10B	36.6	0-200	2400
Circuit Bd. L	FR	11	34.8	0-200, 255-495, 626-1800	1900

[a]Z-configuration
[b]Vertical configuration
[c]Jacketing only, with an 8 mm O.D. copper tubing insert replacing the contents

TABLE 7 Summary of Furniture Calorimeter Results.

Product		TV NFR	TV FR	TV FR	Bus. Machine NFR	Bus. Machine FR	Chair NFR	Chair NFR	Chair FR	Cable (vertical configuration) NFR	Cable (vertical configuration) FR	Cable (vertical configuration) FR	Cable (Z configuration) NFR	Cable (Z configuration) FR	Circuit Board NFR	Circuit Board FR
Specimen Code		H	G	G	F	A	T	T	S	D	K	Jacket	D	K	C	L
Test No.		3	4	15	1	2	16	18	17	19	20	21	5	6	10B	11
Total Mass	kg	3.7	3.7	3.7	3.5	3.5	5.5	5.3	11.9	11.35	11.52	3.5	17.5	18.2		
Combustible Mass	kg	3.7	3.7	3.7	3.5	3.5	5.5	5.3	11.9	6.24	6.45	3.5	9.6	10.2	36.6	34.8
Mass Loss	kg	3.6	2.1	2.0	3.2	2.5	5.2	5.1	**	4.6	1.60	2.0	3.5	2.2	13.4	1.9
Peak Heat Release Rate	kW	515	180	175	560	380	1160	1205	50*	400	75*	140	245	130	205	100*
(Time of Occurrence)	s	139	216	88	138	186	218	208	209	858	1208	265	839	1402	396	1863
Total Heat	MJ	83	40	40	75	69	136	135	*	188	*	67	124	75	238	*
Average Heat of Combustion	MJ/kg	23	20	20	24	28	26	27	*,**	41	*	34	35	34	18	*
Average CO	kg/kg	0.12	0.48	0.26	0.13	0.29	0.01	0.01	**	0.12	0.10	0.13	0.25	0.30	0.10	0.10
Average CO_2	kg/kg	1.39	0.72	0.75	1.61	1.45	1.88	1.89	**	1.61	1.04	1.48	1.89	0.70	1.71	1.36
Average HCl	kg/kg									0.12[a]	0.13[a]					
Average HBr	kg/kg			0.08[a]												
Average HCN	kg/kg						0.001[a]									
Average Smoke Extinction Area	m²/kg	1320	2690	2910	1145	1280	210	165	180	280	235	560	375	545	285	115

*Not reliable. Specimen heat release rate accuracy of ± 25 kW
**Not reliable. Specimen weight loss comparable to noise level of instrumentation
[a]Determined by ion chromatography

TABLE 8 Comparison of Cone Calorimeter Versus Furniture Calorimeter Data.

Sample	NFR /FR	Pk RHR (kW/m^2)[a]		Δh_c (MJ/kg)		% burned		Avg. CO (kg/kg)		Avg. CO$_2$ (kg/kg)	
		Cone	Furn.[b]	Cone	Furn.[b]	Cone	Furn.[b]	Cone	Furn.[b]	Cone	Furn.[b]
TV Cabinet H	NFR	970	520	30	23	99	97	0.015	0.12	2.28	1.39
TV Cabinet G	FR	340	180	12	20	98	57	0.109	0.37	0.67	0.74
Bus. Machine F	NFR	650	560	30	24	88	91	0.037	0.13	2.21	1.61
Bus. Machine A	FR	380	380	21	28	81	71	0.055	0.29	1.60	1.45
Chair T	NFR	470	1180	27	27	89	96	0.020	0.01	1.62	1.89
Chair S	FR	290	50	18	c,d	67	d	0.051	d	0.96	d
Cable D	NFR	360	400	28	41	35	41	0.041	0.12	1.77	1.61
Cable K	FR	380	80	23	c	33	14	0.060	0.10	1.34	1.04
Circuit Bd. C	NFR	250	210	21	18	28	37	0.014	0.10	2.07	1.71
Circuit Bd. L	FR	100	100	13	c	36	5	0.103	0.10	0.87	1.36

Sample	NFR /FR	HCl (kg/kg)		HBr (kg/kg)		HCN (kg/kg)		Smoke (m^2/kg)	
		Cone	Furn.[b]	Cone	Furn.[b]	Cone	Furn.[b]	Cone	Furn.[b]
TV Cabinet H	NFR	NM	NM	NM	NM	NM	NM	1010	1320
TV Cabinet G	FR	NM	NM	0.07	0.08	NM	NM	1880	2800
Bus. Machine F	NFR	NM	NM	NM	NM	NM	NM	1710	1150
Bus. Machine A	FR	NM	NM	NM	NM	NM	NM	1660	1280
Chair T	NFR	NM	NM	NM	NM	0.002	0.001	410	190
Chair S	FR	0.02	NM	TR	NM	0.005	d	480	180
Cable D	NFR	0.11	0.12	NM	NM	NM	NM	1010	280
Cable K	FR	0.13	0.13	NM	NM	NM	NM	880	240
Circuit Bd. C	NFR	NM	NM	NM	NM	NM	NM	560	290
Circuit Bd. L	FR	NM	NM	0.02	d	NM	NM	400	120

NOTE: All Cone Calorimeter data refer to 30 kW/m^2 irradiance tests.

[a] Values for the Cone Calorimeter refer to peak q''_{os} (kW/m^2); values for the Furniture Calorimeter refer to peak q (kW).
[b] Values obtained from the following Furniture Calorimeter tests:
    H: 3
    G: 4,15
    F: 1
    A: 2
    T: 16,18
    S: 17
    D: 19
    K: 20
    C: 10B
    L: 11

[c] Not reliable. Specimen heat release rate accuracy of ± 25 kW
[d] Not reliable. Specimen weight loss comparable to noise level of instrumentation
NM Not measured
TR Trace

TABLE 9 N-gas Model Toxicity Procedure.

1. Material is thermally decomposed.
2. Concentrations of major toxicants are measured.
3. LC$_{50}$ is predicted based on experimental data of toxic interactions.
4. Rats are exposed to predicted LC$_{50}$.
5. If some percentage die, can attribute cause to major toxicants.
6. Death of all animals indicates other gases are contributing to toxicity.
   A. Determine the LC$_{50}$ value.
   B. Low value means extremely toxic.
   C. Need to include other toxic gases means unusually toxic.

Advantages

1. Rapid screening test
2. Economical
3. Minimizes the use of animals
4. More information than just toxic potency

TABLE 10 Chemical and Toxicological Results from Materials Decomposed in the Flaming Mode.

Specimen	NFR/FR	Material	Expt. Temp. (°C)	Mass Load (mg/l)	Mass Cons. (mg/l)	CO (ppm)	CO_2 (ppm)	HCN (ppm)	O_2 (%)	HBr (ppm)	N-Gas Prediction Within	N-Gas Prediction Post	Deaths Within Exp.	Deaths Within & Post	Latest Day of Death
TV Cabinet H	NFR	polystyrene	550[b]	40	40	2740	40600	NDT[c]	15.3	ND[d]	0.99	—	—	—	—
				30	30	2230	39200	ND	15.8	ND	0.83	—	0/6	0/6	—
				40	40	2560	39800	ND	15.4	ND	0.93	—	1/6	1/6	0
TV Cabinet G	FR	polystyrene	525[b]	40[e]	40	3500	9020	NDT	19.7	ND	0.65	—	—	—	—
				40[f]	40	7400	20800	ND	17.9	ND	1.54	—	—	—	—
				20[f]	20	2740	12300	ND	19.4	ND	0.56	—	0/6	0/6	—
				30[f]	30	3840	14000	ND	18.9	163	0.79	—	0/6	2/6	1
				40[f]	40	6570	22200	ND	17.8	ND	1.42	—	6/6	6/6	0
Bus. Machine F	NFR	polyphenylene oxide	600	40	38	5100	38600	ND	15.7	ND	1.47	—	—	—	—
				32	30	3800	33600	ND	16.5	ND	1.08	—	0/6	0/6	—
				35	33	4700	37800	ND	15.8	ND	1.36	—	6/6	6/6	0
Bus. Machine A	FR	polyphenylene oxide	575	40	38	8100	30400	ND	16.7	ND	1.91	—	—	—	—
				25	24	5000	23400	ND	17.9	ND	1.13	—	4/6	4/6	0
Chair foam T	NFR	polyurethane	425[g]	40	35	680	36800	16	16.2	ND	0.55	0.60	—	—	—
				40	35	690	39400	14	16.1	ND	0.56	0.60	0/6	0/6	—
Chair foam S	FR	polyurethane	525	40	26	2800	13900	38[h]	19.2	ND	0.83	0.94	—	—	—
			550	40	26	2800	14100	24[h]	19.2	ND	0.74	0.82	—	—	—
				40	24	3100	13800	17[h]	19.2	ND	0.75	0.80	—	—	—
				40	26	3600	20200	45[h]	18.4	ND	1.10	1.23	—	—	—
				30	20	2700	18400	68[h]	18.7	ND	1.05	1.25	0/6	0/6	—
				35	23	3000	19400	60[h]	18.7	ND	1.06	1.23	0/6	2/6	1
Cable D	NFR	EVA	450	40	34	1340	38500	ND	15.3	ND	0.66	—	—	—	—
				40	34	1270	41600	ND	14.8	ND	0.69	—	0/6	0/6	—
				80	68	4420	66600	ND	12.9	ND	1.47	—	6/6	6/6	0
Cable K	FR	EVA	450	40	32	3080	17200	ND	18.4	ND	0.71	—	—	—	—
				20	17	1650	13500	ND	18.9	ND	0.41	—	0/6	1/6	19
				40	33	3150	23500	ND	17.6	ND	0.81	—	1/6	6/6	1
Circuit Bd. C	NFR	polyester	550	41	12	600	12700	ND	19.7	ND	0.18	—	—	—	—
				41	12	600	15600	ND	19.3	ND	0.21	—	0/6	0/6	—
				165	48	3010	45700	ND	15.5	ND	1.08	—	0/6	0/6	—
				200	58	3870	50600	ND	15.0	ND	1.38	—	6/6	6/6	0
Circuit Bd. L	FR	polyester	625	40	19	2030	10800	ND	19.7	ND	0.42	—	—	—	—
				41	19	1850	9800	ND	19.7	ND	0.39	—	0/6	0/6	—
				60	28	3000	18200	ND	18.4	45	0.70	—	0/6	0/6	—
				85	36	3240	20200	ND	18.3	39	0.76	—	0/6	0/6	—
				100	45	4820	25000	ND	17.6	57	1.13	—	6/6	6/6	0

[a]Average concentration over 30-minute exposure.
[b]In non-flaming mode, no deaths occurred.
[c]NDT = Not detected.
[d]N = Not determined.
[e]Sparker turned off after first flicker.
[f]Sparker on until flaming stopped.
[g]In non-flaming mode at 40 mg/l, 3/6 animals died on days 7, 11, and 14.
[h]HCN results were variable.

N-Gas prediction based on equation:

$$\frac{m[CO]}{[CO_2]-b} + \frac{[HCN]}{X} + \frac{21-[O_2]}{21-5.4}$$

where m = slope of LC_{50} line of CO in the presence of CO_2 and b = the y intercept of this line. Values used for m (−18.4 or 22.7 when CO_2 is below or above 5%, respectively) and b (122,000 or −39,000 when CO_2 is below or above 5%, respectively) were based on data obtained after 1/1986. X = LC_{50} value of HCN which is 160 ppm for deaths during the 30-min exposures and 110 ppm when deaths occur within 24 hours post-exposure. 5.4% is the 1986 30-min LC_{50} value for O^2.

TABLE 11 Summary of LC_{50} Values and 4-Gas Model Prediction Values Determined in the Small-Scale Toxicity Tests.

Specimen	NFR /FR	Material	Estimated LC_{50} Mass Loading (mg/l)		Estimated LC_{50} Mass Consumed (mg/l)		N-Gas Prediction Model		Other Gases Needed	Extreme Toxic Potency
			Within Exp.	Within & Post-Exp.	Within Exp.	Within & Post-Exp.	Within Exp.	Within & Post-Exp.		
TV Cabinet H	NFR	polystyrene	40	40	40	40	1.0	1.0	No	No
TV Cabinet G	FR	polystyrene	30-40	30	30-40	30	0.79-1.42	0.80	?	No
Bus. Machine F	NFR	polyphenylene oxide	32-35	32-35	30-33	30-33	1.08-1.36	1.08-1.36	No	No
Bus. Machine A	FR	polyphenylene oxide	25	25	24	24	1.13	1.13	No	No
Chair T (foam)	NFR	polyurethane	>40	>40	>35	>35	>0.56	>0.60	?	No
Chair S (foam)	FR	polyurethane	>35	≈35	>23	≈23	>1.06	≈1.23	No	No
Cable D	NFR	EVA	40-80	40-80	34-68	34-68	0.69-1.47	0.69-1.47	?	No
Cable K	FR	EVA	>40	20-40	>33	17-33	>0.81	0.41-0.81	Yes	No
Circuit Bd. C	NFR	polyester	165-200	165-200	48-58	48-58	1.08-1.38	1.08-1.38	No	No
Circuit Bd. L	FR	polyester	85-100	85-100	36-45	36-45	0.76-1.13	0.76-1.13	?	No

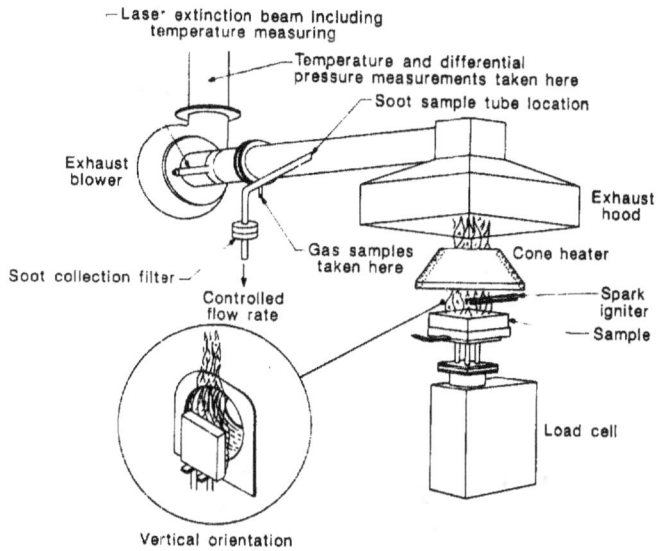

Figure 6. Conceptual view of the Cone Calorimeter.

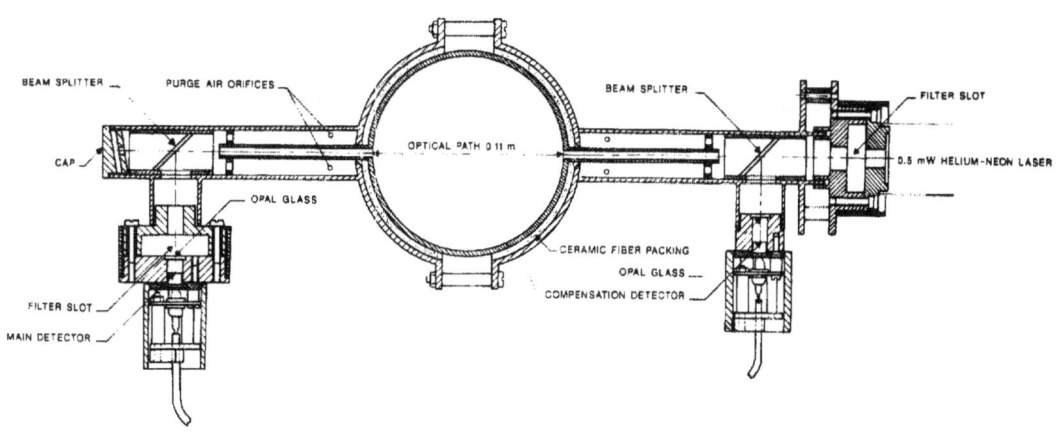

Figure 7. Smoke measurement system used on the Cone Calorimeter.

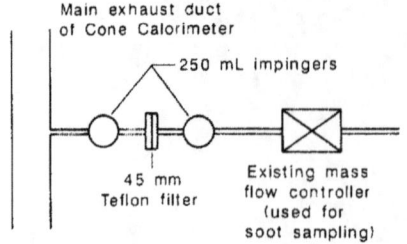

Figure 8. Impinger gas sampling in the Cone Calorimeter.

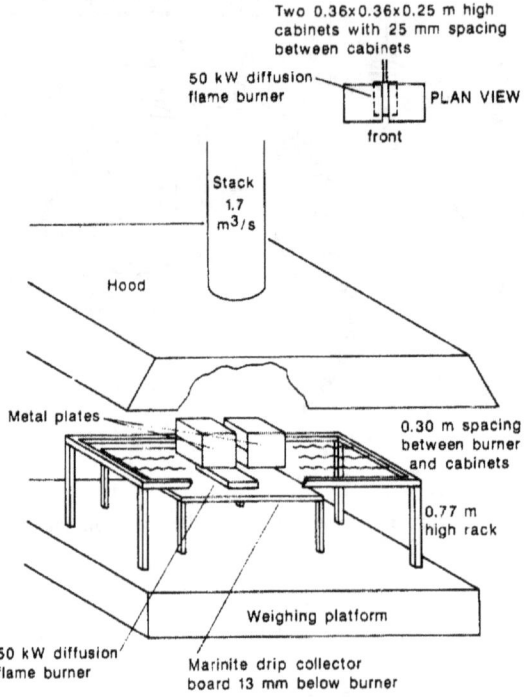

Figure 9. Furniture Calorimeter test of business machines and TV cabinets.

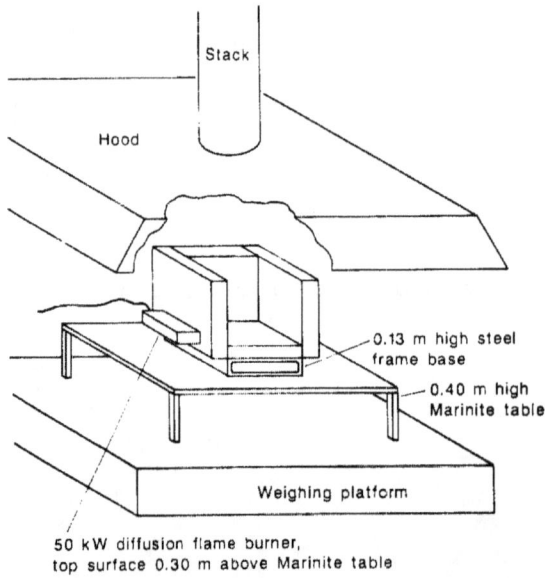

Figure 10. Furniture Calorimeter test of chairs.

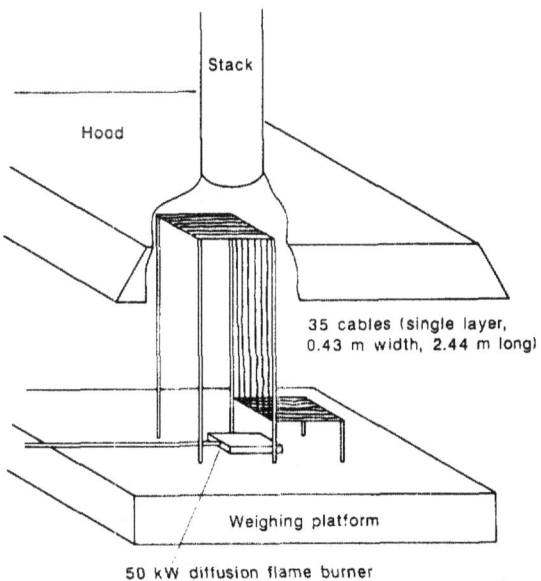

Figure 11. Furniture Calorimeter test of cables (Z-configuration).

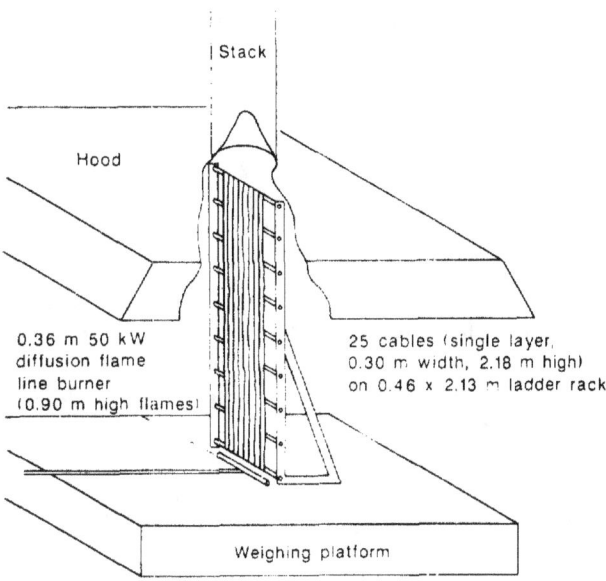

Figure 12. Furniture Calorimeter test of cables (vertical array).

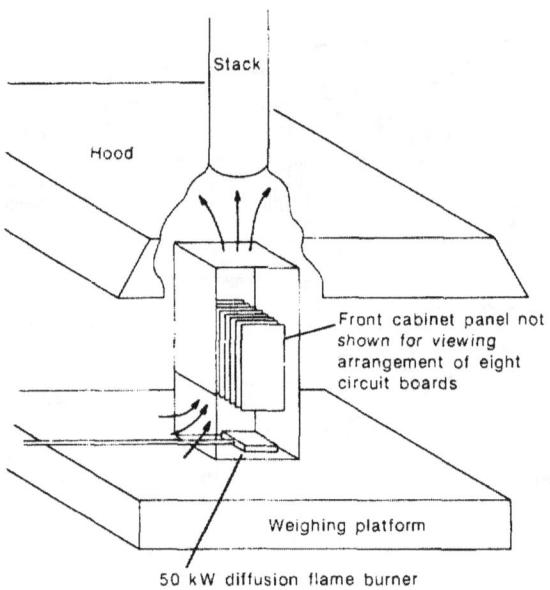

Figure 13. Furniture Calorimeter test of circuit boards.

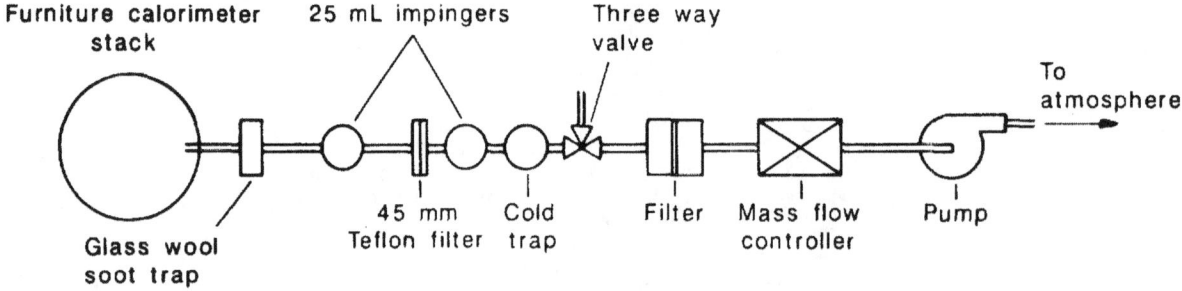

Figure 14. Impinger gas sampling in the Furniture Calorimeter.

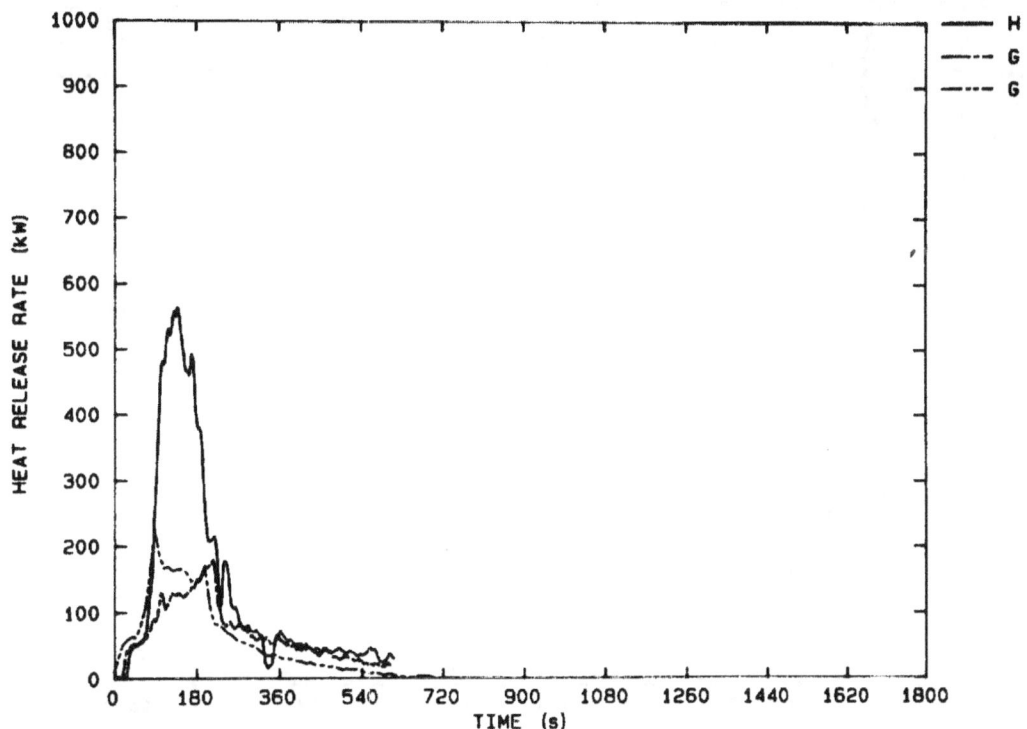

Figure 15. Rate of heat release for TV cabinets measured in the Furniture Calorimeter.

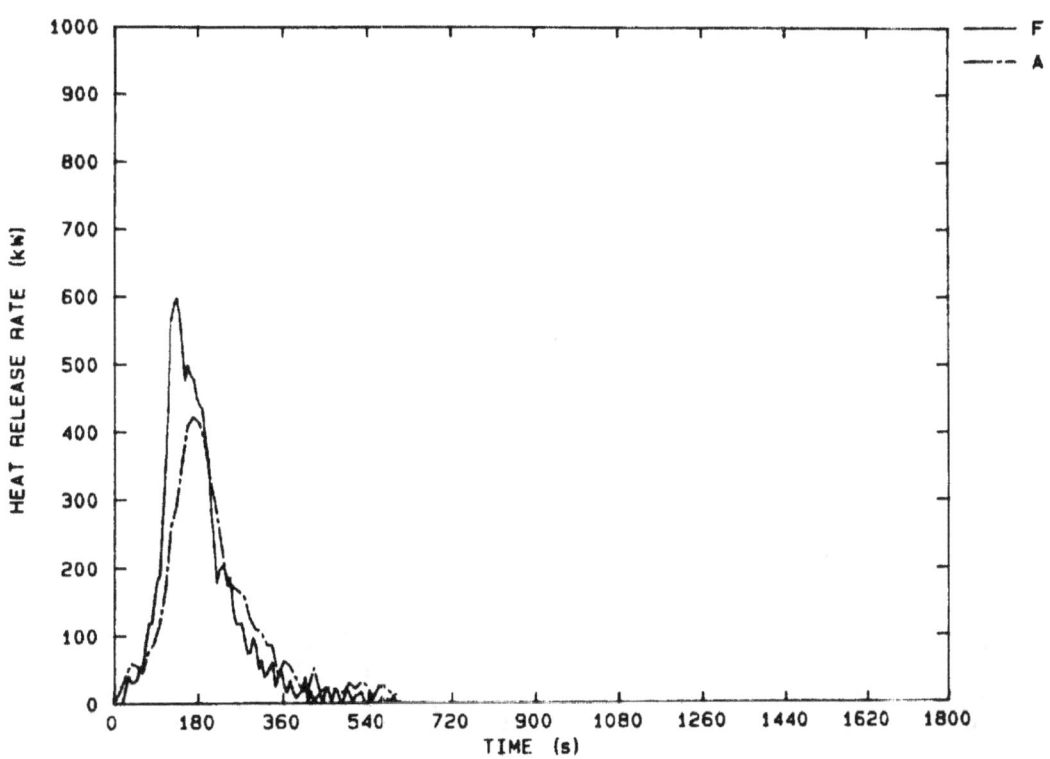

Figure 16. Rate of heat release for business machine housings measured in the Furniture Calorimeter.

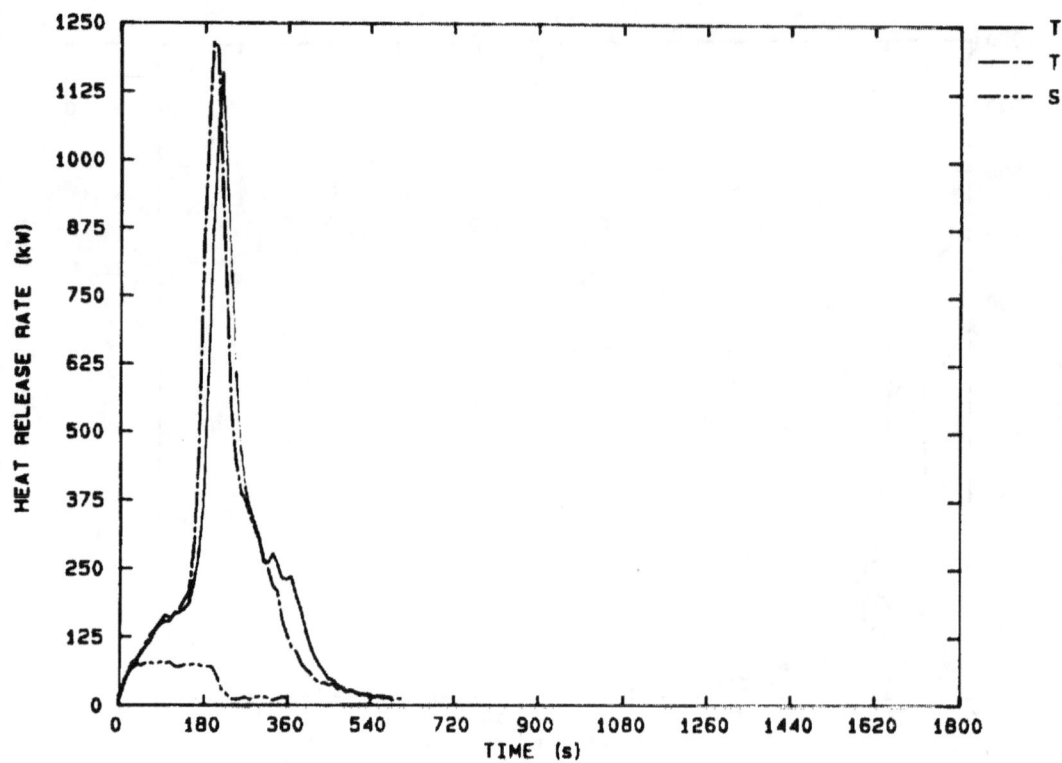

Figure 17. Rate of heat release for upholstered chairs measured in the Furniture Calorimeter.

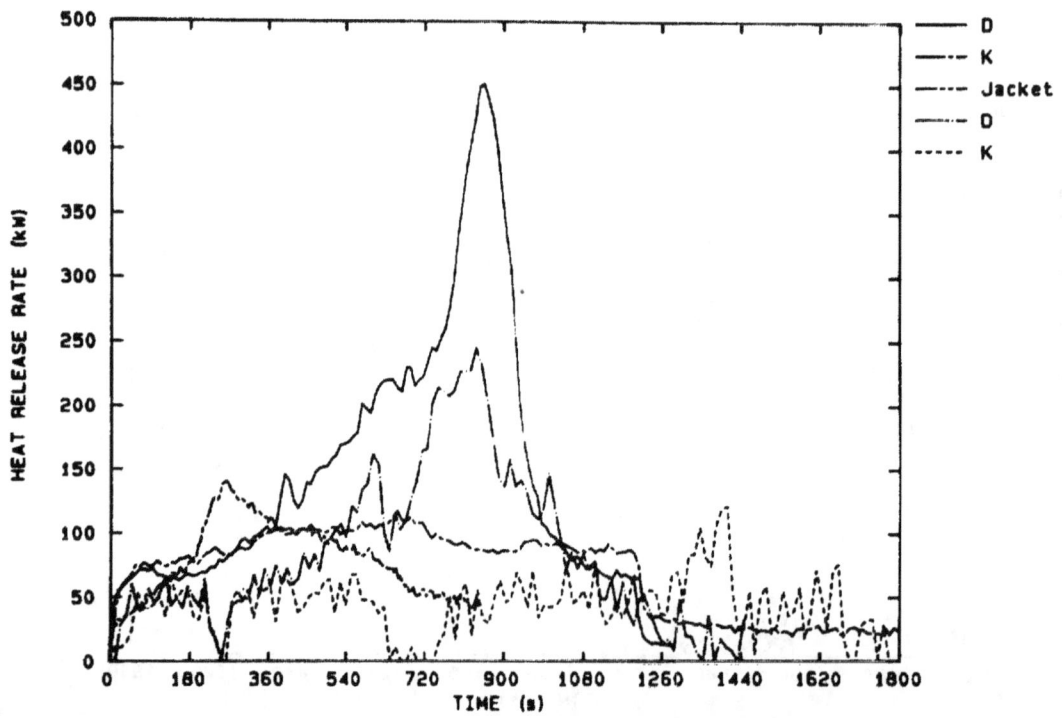

Figure 18. Rate of heat release for cables measured in the Furniture Calorimeter.

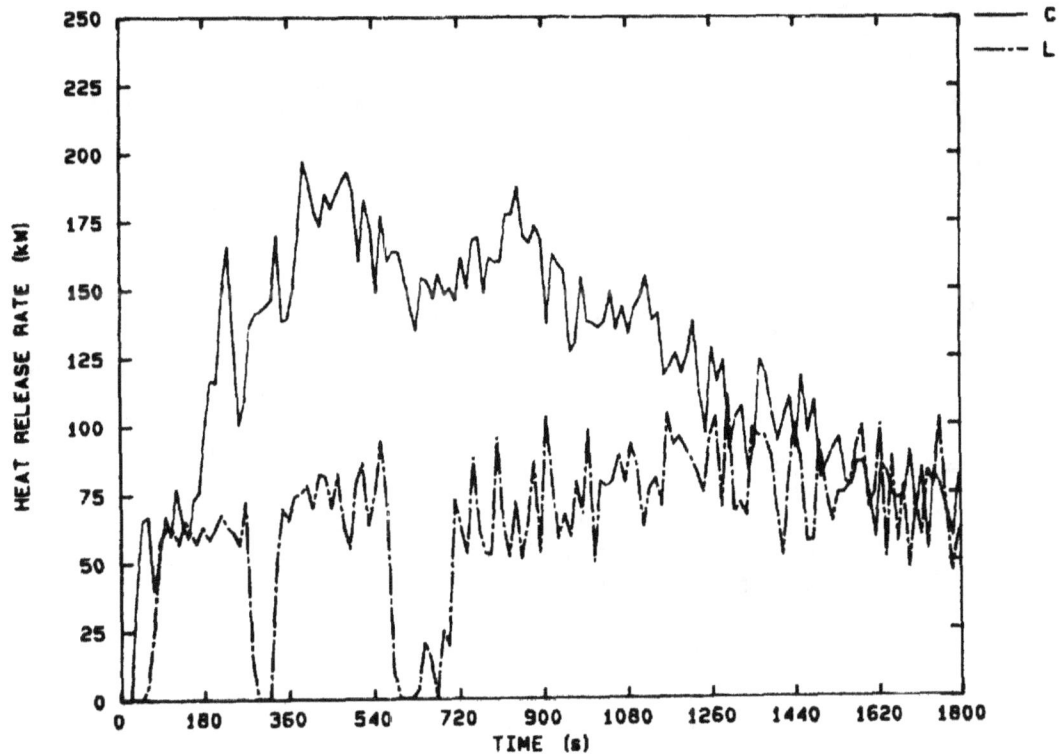

Figure 19. Rate of heat release for circuit boards measured in the Furniture Calorimeter.

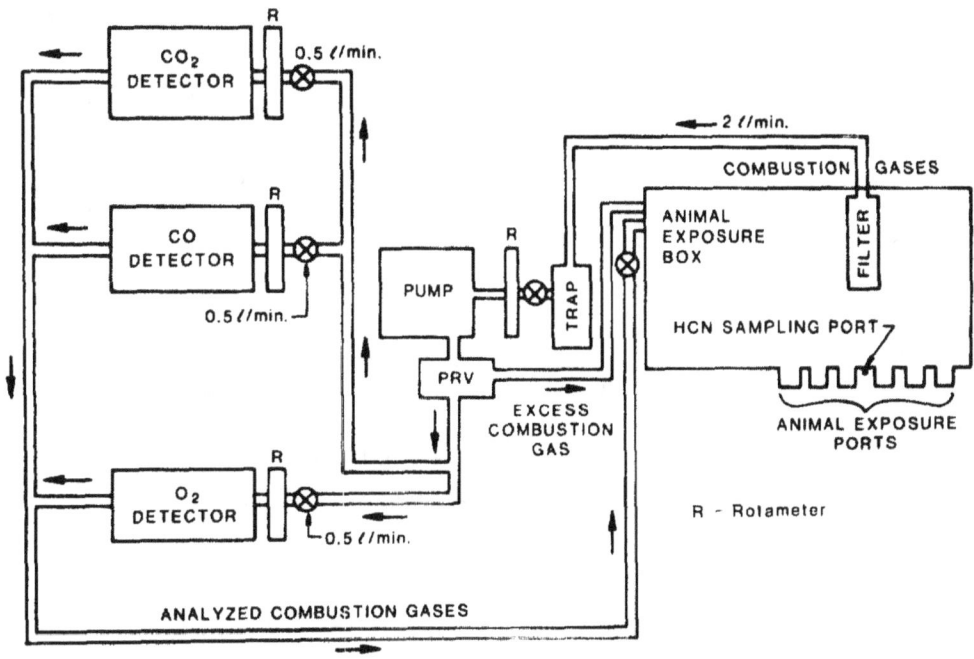

Figure 20. Schematic of gas analysis system used in the small-scale toxicity test.

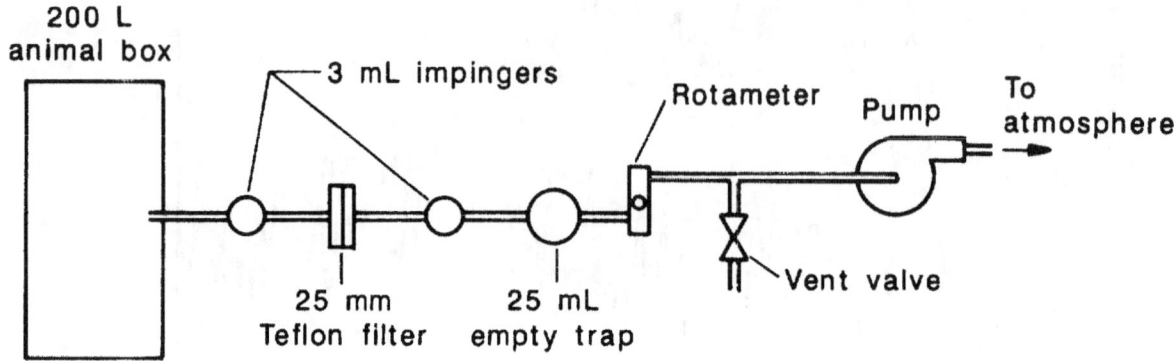

Figure 21. Impinger gas sampling in the small-scale toxicity test.

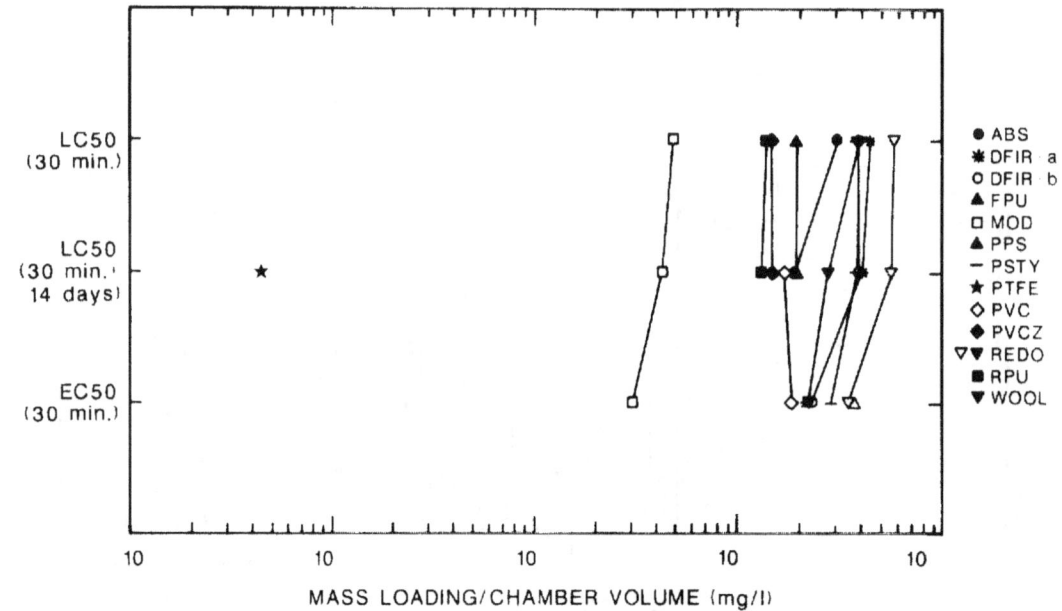

Figure 22. Comparison of materials by their EC_{50}, LC_{50} (30 min.) and LC_{50} (30 min. + 14 days) after flaming decomposition.

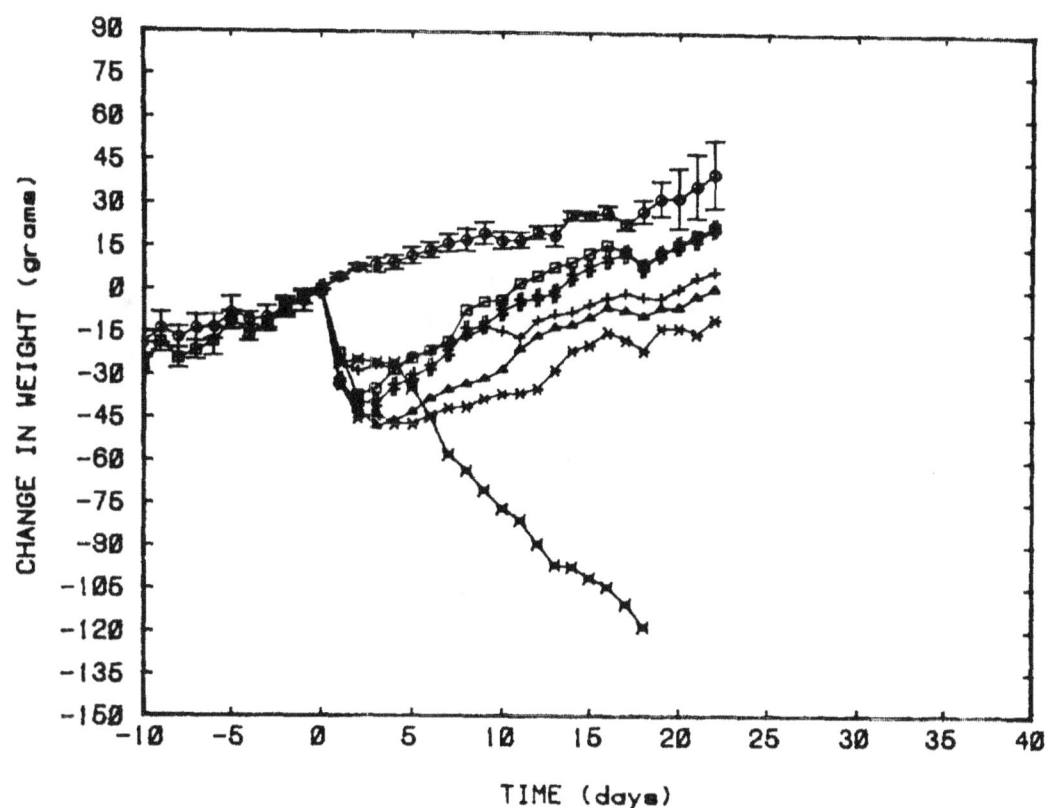

Figure 23. Animal weight change following a 30 minute exposure to the flaming decomposition products from a mass loading of 20 mg/l of specimen K.

4

Model Simulations of the Large-Scale Room/Corridor/Room Fires

DUE to the high cost of full-scale testing, considerable care must be taken in the design of appropriate test scenarios to be conducted. To determine the suitability of any chosen large-scale test arrangements, fires involving combinations of the test products were simulated on the computer using the FAST model to predict conditions in rooms arranged in different configurations [21]. With the use of such a predictive model, the number of large-scale tests can be reduced to a minimum by only conducting tests in the appropriate configurations. As described below, two different potential room/corridor/room configurations and three doorway opening sizes to the outside were explored. These computer simulations would not only be used to visualize the impact of the FR additives on the fire development in the large scale, but also proved essential in rejecting inadequate test geometries.

4.1 PRODUCT ARRANGEMENT A AND ROOM CORRIDOR CONFIGURATION A

Initially, the computer simulations were done for the product arrangement A shown in Figure 24 with the room-corridor configuration A indicated in Figure 25. A 50 kW diffusion burner, using natural gas and positioned against one side of the chair, served as the fire initiation source. An auxiliary burner in one back corner of the burn room served as a possible fire source prior to ignition of the products. A 2 m wide by 1 m high doorway was chosen between the corridor and the outside. The low height was intended to help trap combustion gases in the corridor and to force these gases into the side target room. The broad width was used to ensure adequate fresh air for combustion.

4.1.1 NFR PRODUCTS

For the NFR items arranged as in Figure 24, data from the Furniture Calorimeter predicted that chair T would ignite upon exposure to the 50 kW burner flames, and flame spread across the burning chair would reach the cables D and circuit board C by 175 s. At about 200 s later (burn time of 375 s), the circuit board and cables would ignite and flames would contact the adjacent television cabinets H. After 75 s, item H would ignite and become fully involved by another 125 s (burn time of 575 s), by which time flames would be contacting the business machine cabinets F. Item F would ignite in another 75 s (burn time of 650 s). The FAST computer fire model [21] uses mass loss rates as the basic input data describing the fire. For the NFR items arranged as in Figure 24, the presumed mass loss rates for this case are shown in Figure 26. A composite mass loss record is shown in Figure 27. This consumption history was used in the FAST model program [21] to predict the fire development in the room/corridor/room spaces.

4.1.2 FR PRODUCTS WITH AUXILIARY BURNER

With the FR products, the Furniture Calorimeter tests showed that chair S would not support continued flame spread under free-burn, ambient conditions. Data from the Cone Calorimeter indicated that the flame spread threshold for chair S was about 10 kW/m², and that only about 20% of chair S would be consumed under that flux environment. For the model simulation it was desired to set the heat release rate from the auxiliary burner such that flame spread would be sustained in chair S. Such an auxiliary source of heat might be expected to occur in any number of realistic scenarios. If such additional heating were not included, however, it is likely that the large-scale results could show no sustained flame propagation at all. A finding of this nature would not, then, be taken as a conservative, useful scenario.

Based on earlier data on fire buildup in well-insulated rooms [22] and on the fact that the actual burn room surfaces are not as well insulated, a total heat source of 300 to 350 kW in the room was estimated to be needed for achieving a 10 kW/m² exposure in the lower part of the room and for sustaining flame spread over chair S. To verify this, a preliminary experiment was performed in the burn room with a gas burner in one back corner operating at 300 kW, and the 50 kW ignition source burning under one edge of a single, horizontal chair S cushion. It was observed that flame spread was sustained and took 240 s to traverse the top surface of the cushion.

In the computer simulation fire, a 250 kW rate was used for the auxiliary burner, a 50 kW rate was used for the ignition burner, and 20% (or 2.38 kg) of chair S was assumed to be consumed based on data from the Cone Calorimeter. It was assumed further that its burning history curve would have a similar shape as that for the NFR chair T. As for the other FR products, cable K did not burn readily in the Furniture Calorimeter and was expected to act as a fire barrier to the TV cabinets G and the business machine housings A. The circuit board material L was not expected to contribute significantly to the fire. Mass loss rate curves corresponding to these set of assumptions are shown in Figure 28.

4.1.3 MODEL PREDICTIONS FOR CONFIGURATION A

The computer predictions for both NFR and FR scenarios indicated that there might not be enough hot combustion gases entering the side target room, for adequate resolution of the toxicological differences in the combustion gases between burn tests of the NFR versus FR products.

4.2 PRODUCT ARRANGEMENT B AND ROOM/CORRIDOR CONFIGURATION B

To create a scenario where a measurable flow of combustion gases would go to the target room, an alternative geometrical arrangement was devised. This was room/corridor/room configuration B, shown in plan view in Figure 29, in conjunction with the product arrangement B, shown in Figure 30. Two possible doorway openings from the target room to the outside were considered. One case used a typical doorway 0.76 m wide and 2.03 m high. In the other case, an undersized 1.0 m wide 1.07 m high doorway was used to trap more heat and combustion gases in the target room.

Analysis of the Furniture Calorimeter free-burn data for the NFR items indicated that the television cabinets H would be burning at their peak rates at the time when chair T would be entering a vigorous burning period, at about 150 s, leading to probable flashover at about the same time if these items were burning in a room. For calculational purposes, items C, D, and F were assumed to be exposed to flames at 150 s, and behave like their performances in the Furniture Calorimeter. In the computer simulations with this furniture arrangement, production of CO, CO_2, and oxygen depletion, using the data from the Furniture Calorimeter were also included. For simplicity, an average value of 1.6 kg CO_2 per kg product consumed was used. A value of minus 2.1 kg O_2 per kg product consumed (based on the approximate value of 13.1 MJ for each kg O_2 consumed in a fire, and an average Δh_c value of 28 MJ for each kg NFR product burned) was also used. Production of CO_2 from the gas burners was expressed in terms of the per-unit mass of equivalent products burned by the burners based on a heat of combustion of 28 MJ per kg of products. The per-unit-mass generation of CO for each item as measured in the Furniture Calorimeter was used as given in Table 7. In those cases where Furniture Calorimeter data were not available, the corresponding data from the Cone Calorimeter were used. The mass loss and the production of CO for each item were then used to generate composite histories for mass loss and CO production for input to the computer model. Figure 31 shows the mass loss rates presumed to occur for this scenario.

Computer predictions indicated that the undersized target room doorway to the outside resulted in the fire being limited by lack of oxygen early in the run, with predicted concentrations of CO as high as 6900 ppm in the target room. With the larger doorway opening, an upper layer temperature of about 600°C, a value indicative of room flashover, was achieved at about 150 s. Peak CO levels of 3500 ppm in the target room were predicted. Based on the findings that flashover conditions in the burn room and that high concentrations of CO in the target room were theoretically possible, it was decided by FRCA to use arrangement B with a typical doorway in conducting the large-scale experimental work.

4.3 WORST CASE SCENARIO FOR PRODUCT ARRANGEMENT B AND ROOM/CORRIDOR CONFIGURATION B

4.3.1 NFR PRODUCTS

In the predicted scenarios in the preceding section, the burning rates for each individual item were taken directly from the furniture calorimeter data. Under such free-burn conditions, interactions between materials or between materials and heated room surfaces are assumed to be negligible. Using the early time rate of heat release histories from the Furniture Calorimeter tests for chair T and television cabinets H, prior to fire involvement of the remaining products, it can be shown that the summation of the rates from the 50 kW burner, T, and H would be about 0.6 MW at 100 s. This rate, when combustion enhancement from the effect of the enclosure was considered, could lead to flashover. As a worst case scenario and to account for combustion enhancement, it was assumed that flashover would occur and that all of the items would be burning at their peak open burn rates by 100 s. These considerations along with the actual CO and CO_2 production per item (rather than an approximate average 1.6 kg CO_2 per kg product burned) were used in the generation of the composite histories for mass loss, CO and CO_2 for input to the computer model. Figure 32 shows the mass loss rates assumed for this case, when no auxiliary burner is used. Figure 33 gives the estimated mass loss rate for the case of using the auxiliary burner. Note that only the time sequence is different, since the first 300 s comprise only the heating from the auxiliary burner, without the ignition burner being turned on. That is, the ignition burner is turned on at t = 0 in Figure 27, and at t = 300 s in Figure 33. Since worst case burning rates were already assumed for the case without the auxiliary burner, no further rise in burning rates due to the heating of the auxiliary burner was added to produce Figure 33.

4.3.2 FR PRODUCTS

4.3.2.1 Without Auxiliary Burner

With the FR items, the television cabinet G and chair S would ignite upon exposure to the ignition burner flames. However, chair S would not be expected to sustain fire spread beyond the region of contact with the burner flames. Cables K and circuit board L would not burn much, and the cables K would act as a fire barrier to the business machine housings A. Thus, for calculational purposes, only the 50 kW burner and item G would be contributing to the fire buildup. Figure 34 shows the mass loss rates for this case.

4.3.2.2 With Auxiliary Burner

The television cabinets G would ignite almost immediately upon flame contact with the 50 kW burner. Data from the Furniture Calorimeter indicated that these cabinets would produce up to 180 kW. Earlier work indicated that a total of 300 to 350 kW would result in a room thermal flux environment whereby flame spread would be sustained in chair S. Thus, the auxiliary burner need only generate 120 kW to achieve this environment.

In the computer simulation, 120 kW would be prescribed for the auxiliary burner and chair S would be assumed to burn in the manner described earlier under the influence of the auxiliary burner. Based on results of the flame spread experiment with the single chair S cushion discussed earlier, the flames would be expected to traverse the chair surface and to contact the cables K in about 240 s. The Furniture Calorimeter data indicated that the cables K did not burn significantly until after 200 s following flame contact. Therefore, for calculational purposes, the cables were assumed to contribute to the fire at 440 s. An examination of the Furniture Calorimeter mass loss data and heats of combustion from the Cone Calorimeter for the FR products together with the probability that only items G and S would be significantly involved resulted in a composite mass loss history with a heat of combustion of about 13.6 MJ per kg of products burned. This resulted in a value of 0.72 kg CO_2 produced for each equivalent kg of products burned by the gas burner and in minus 1.0 kg O_2 for each kg products consumed in the room fire. These considerations along with the actual CO and CO_2 production per item, as determined with the Furniture Calorimeter, were used as input to the computer model. Figure 35 gives the mass loss rates presumed for this case.

4.3.3 MODEL PREDICTIONS WITH CONFIGURATION B

Model predictions of the average upper layer temperature and the CO concentration in the burn room for the NFR products with and without the auxiliary burner are given in Figures 36 to 39. When 120 kW was prescribed for the auxiliary burner, the time for the rate of heat release of the burning products to reach 0.6 MW under open burn conditions (a value assumed, above, to lead to room flashover when the products were burned in a room) shortened by only 10 s. Thus, the prediction curves for this case, indicated in Figures 31 and 33 can be seen to shift by about the same time interval (taking into due account the 300 s difference in the time axes).

Model predictions as above for the FR products, with and without the auxiliary burner, are indicated in Figures 40 to 43.

4.4 USE OF THE MODEL PREDICTIONS IN THE DESIGN OF LARGE-SCALE TESTS

The model simulations in this report are one of the first use of predictive fire modeling in the design of full-scale fire tests. They proved an invaluable tool in selecting the best physical configuration for the test facility used in the experimental program. Based on these computer predictions for the NFR and FR products using arrangement B and the room-corridor configuration B with the larger doorway opening, FRCA adopted the same arrangement and configuration for the experimental program.

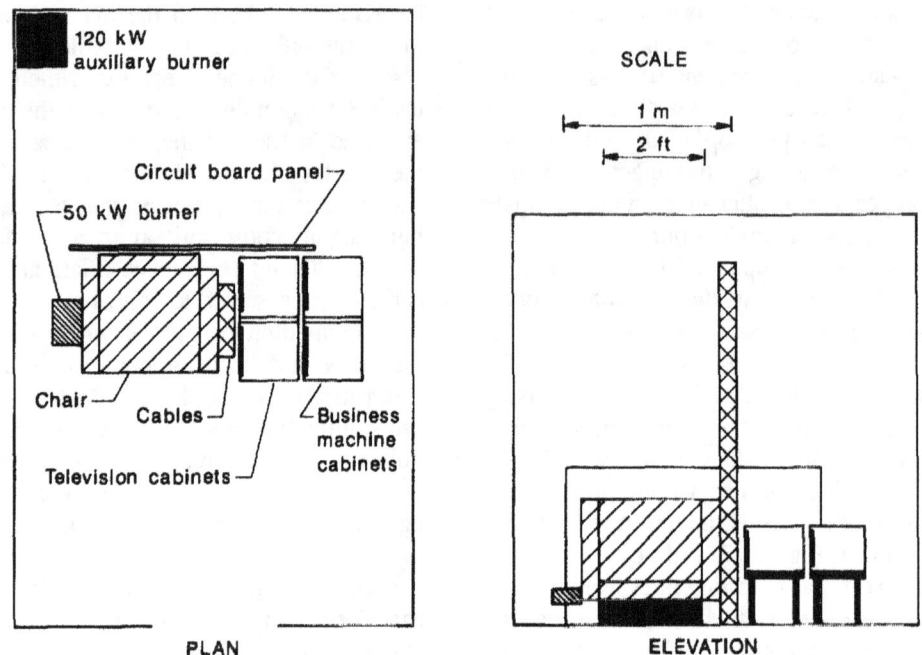

Figure 24. Arrangement A of test items in the burn room.

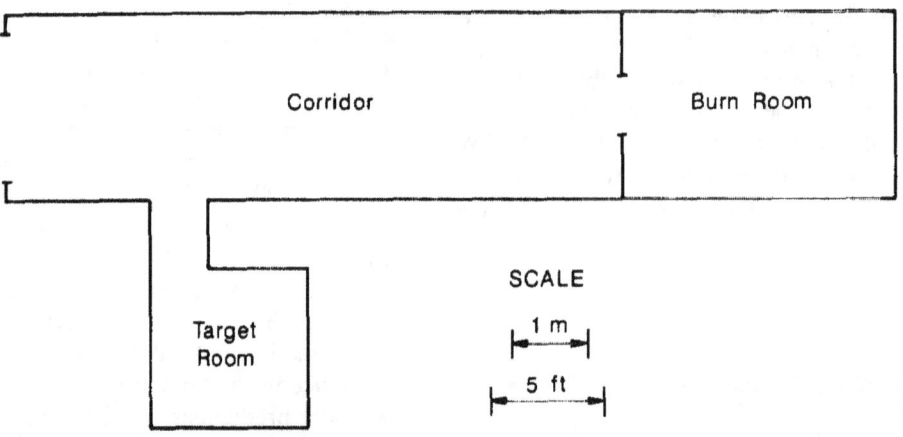

Figure 25. Plan view of large-scale test arrangement A.

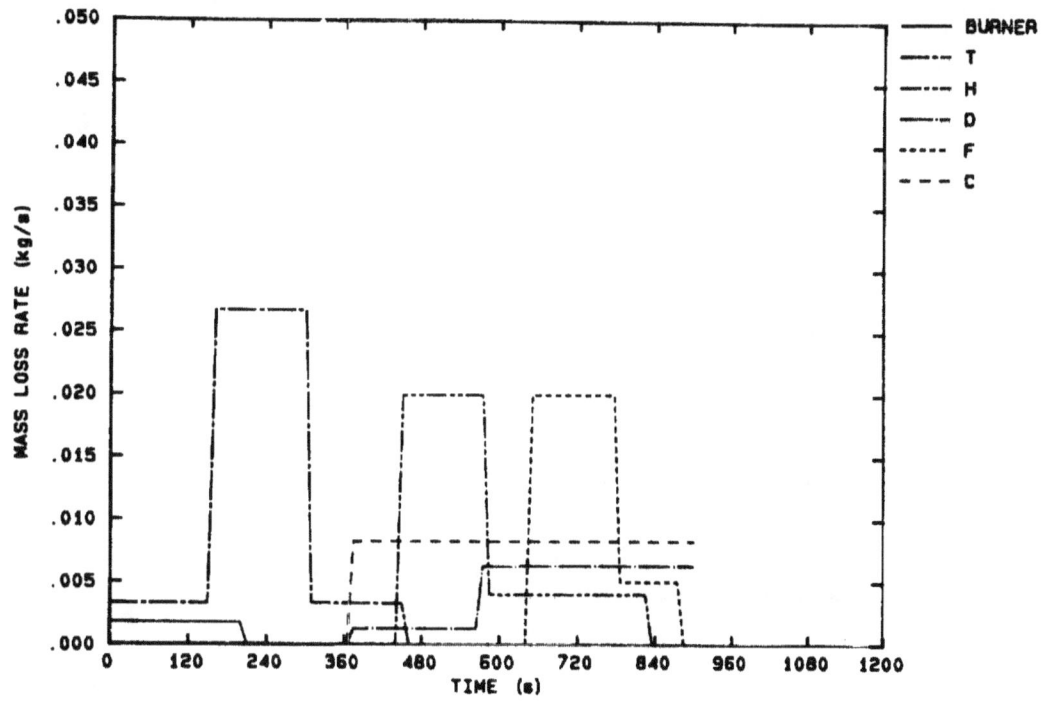

Figure 26. Mass loss rate for NFR furnishing arrangement A and room-corridor configuration A without the auxiliary burner.

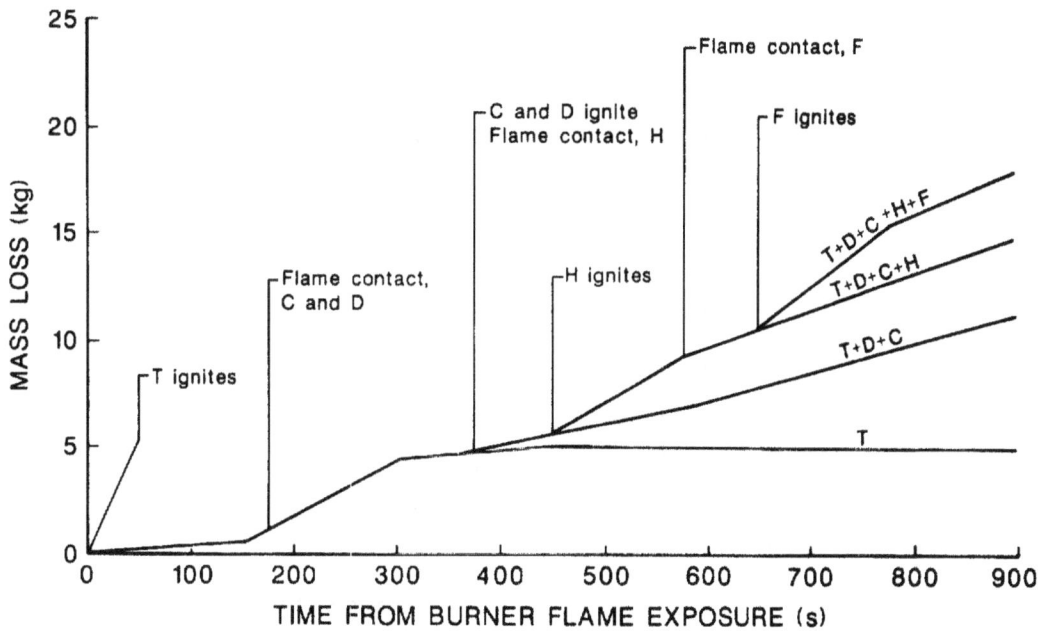

Figure 27. Composite mass loss history for NFR furnishings.

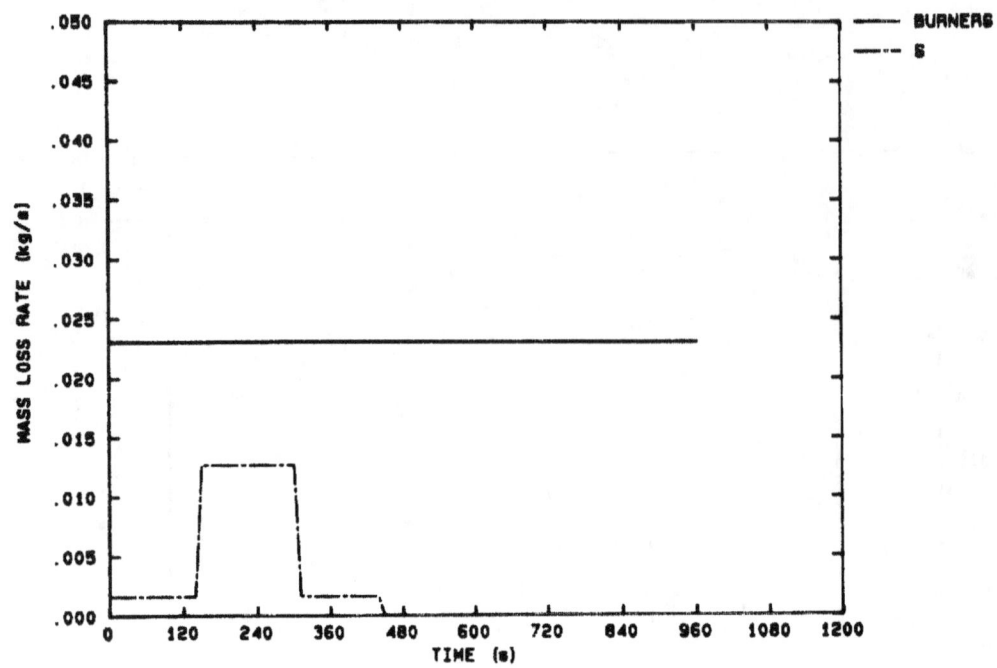

Figure 28. Mass loss rate for FR furnishing arrangement A and room-corridor configuration A with the auxiliary burner.

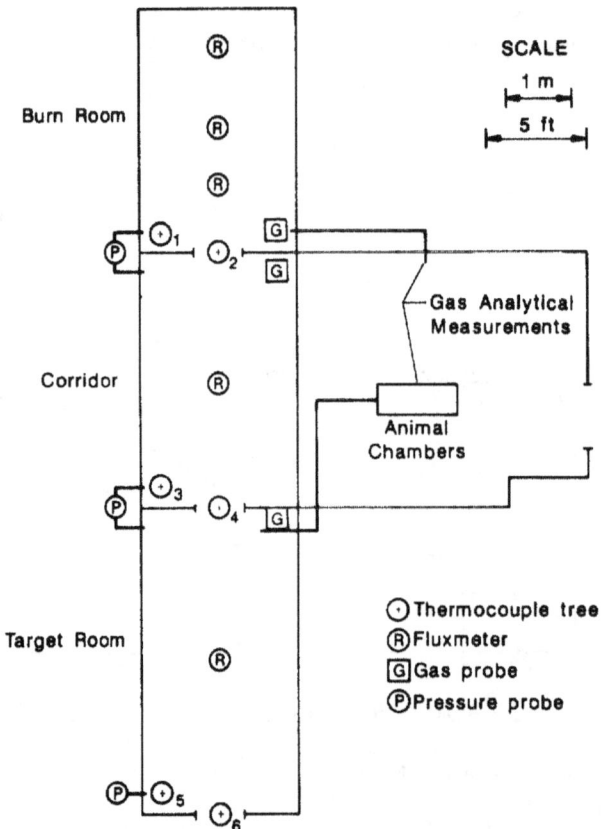

Figure 29. Plan view of large-scale test arrangement B.

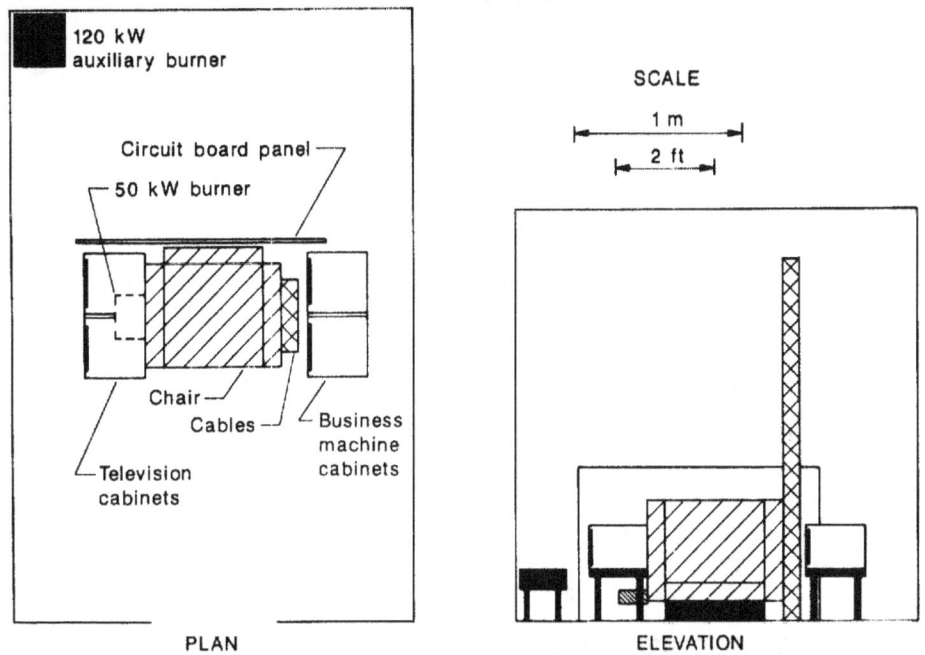

Figure 30. Arrangement B of test items in the burn room.

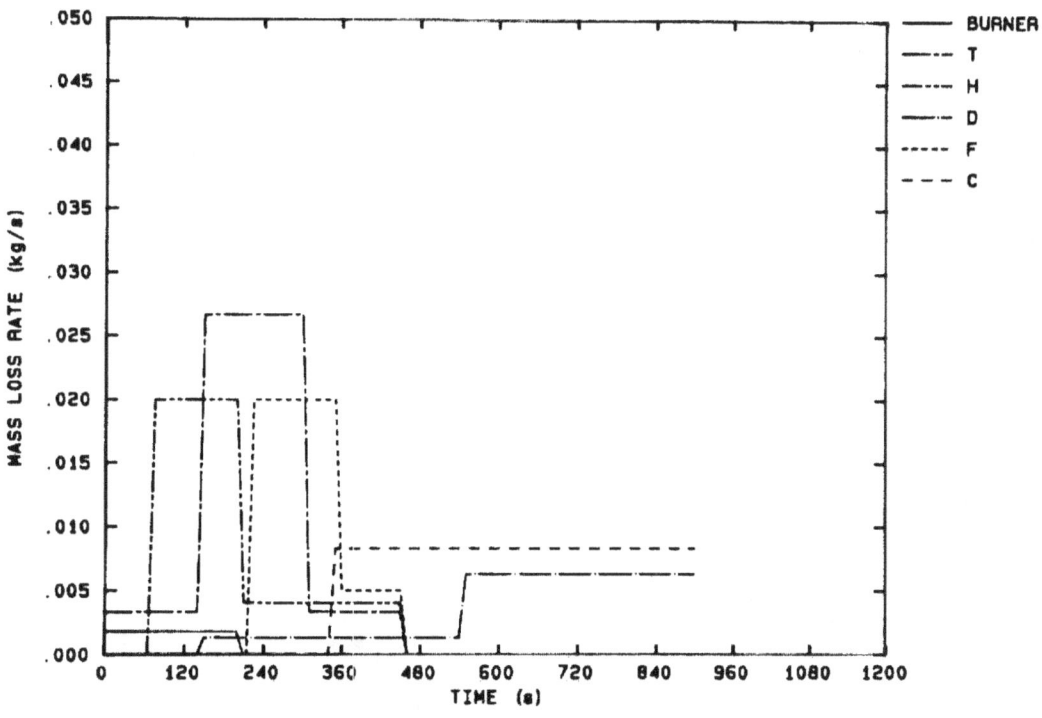

Figure 31. Mass loss rate for NFR furnishing arrangement B and room-corridor configuration B without auxiliary burner.

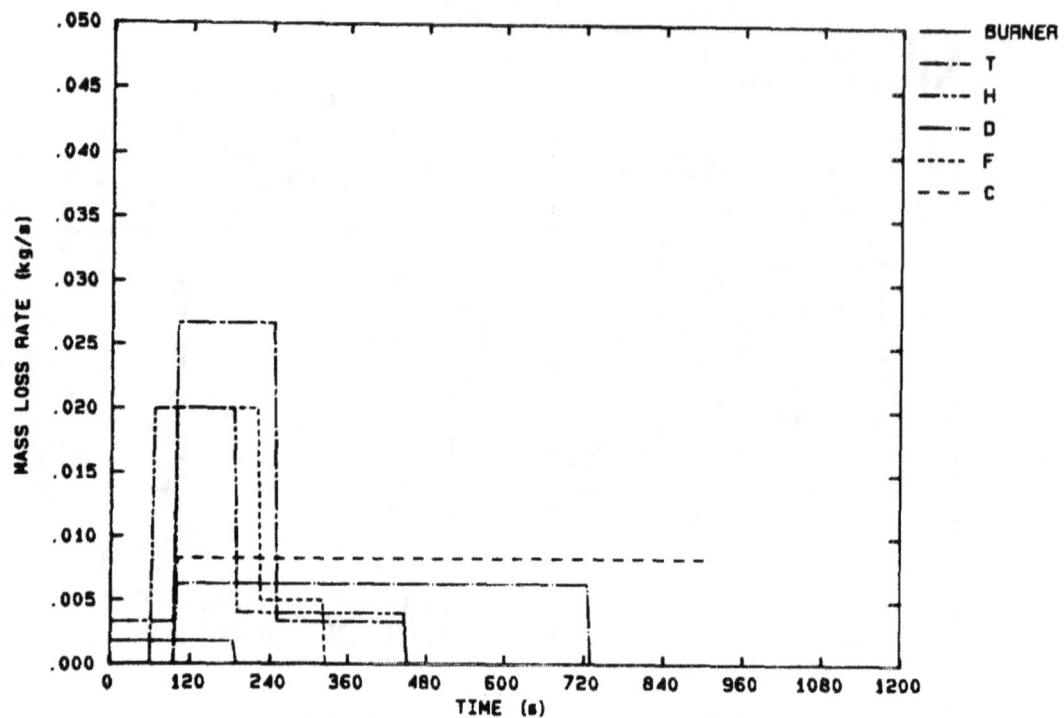

Figure 32. Accelerated mass loss rate for NFR furnishing arrangement B and room-corridor configuration B without the auxiliary burner

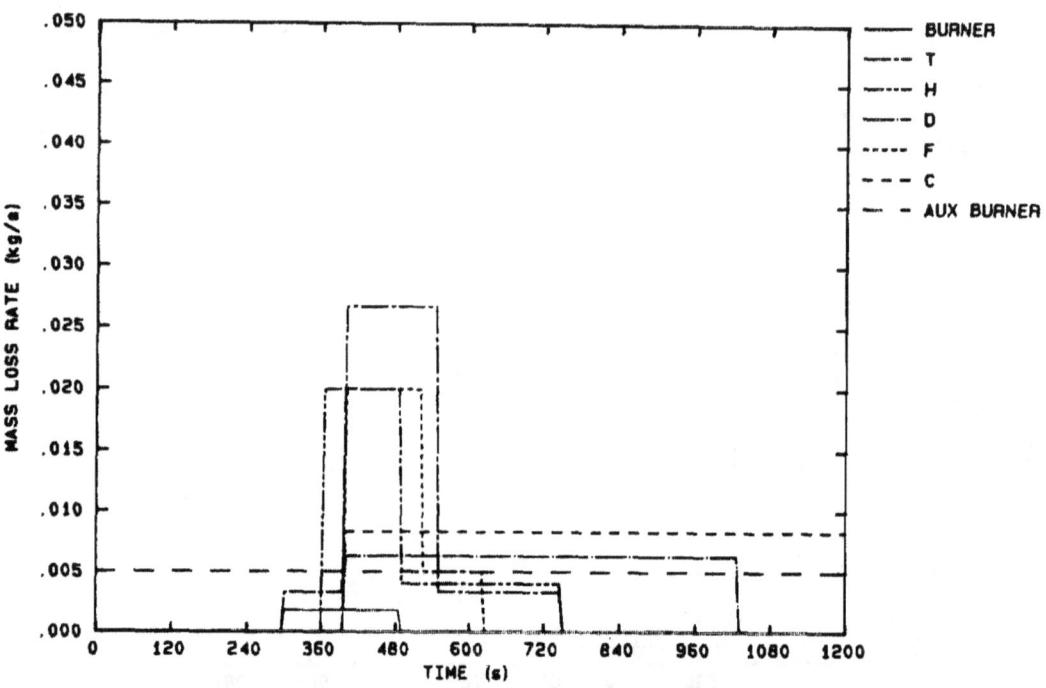

Figure 33. Accelerated mass loss rate for NFR furnishing arrangement B and room-corridor configuration B with the auxiliary burner.

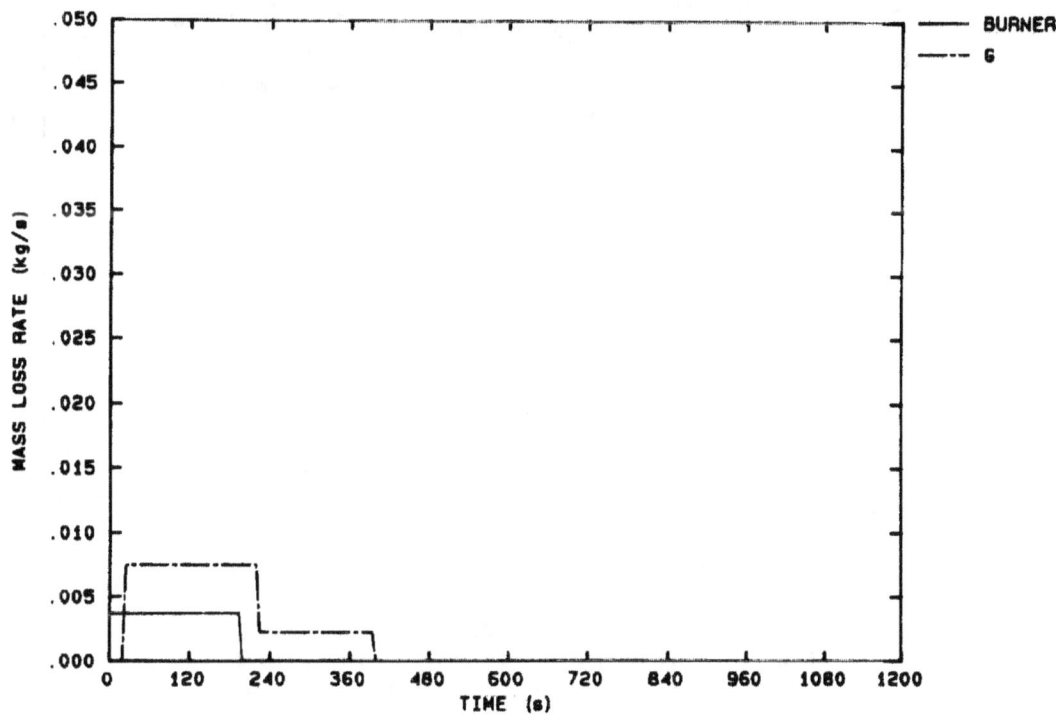

Figure 34. Mass loss rate for FR furnishing arrangement B and room-corridor configuration B without the auxiliary burner.

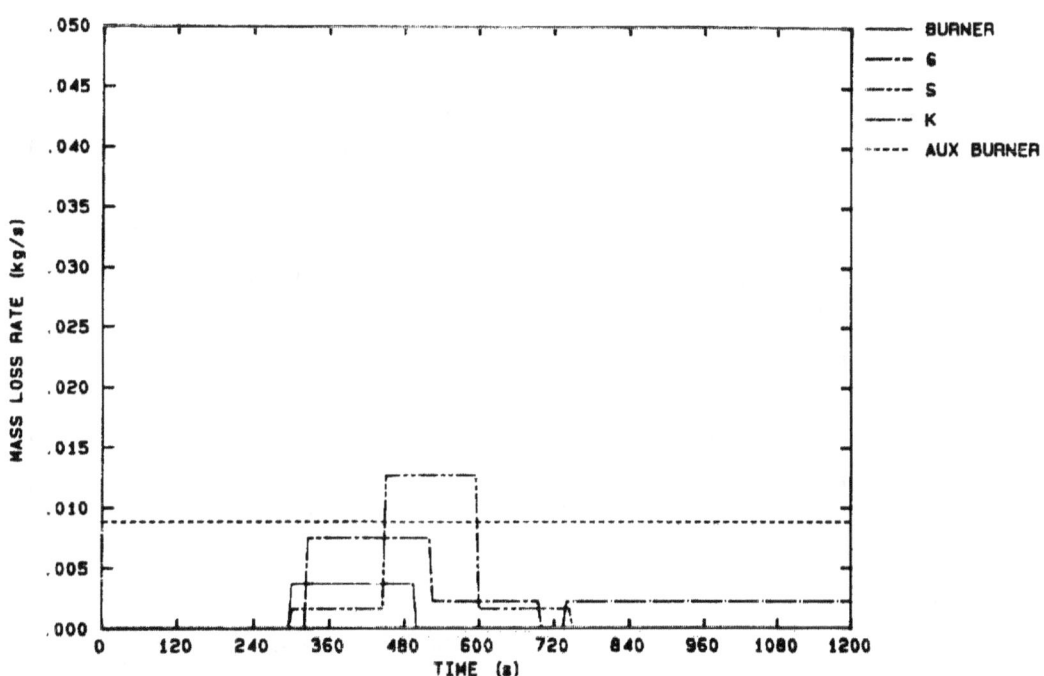

Figure 35. Mass loss rate for FR furnishing arrangement B and room-corridor configuration B with the auxiliary burner.

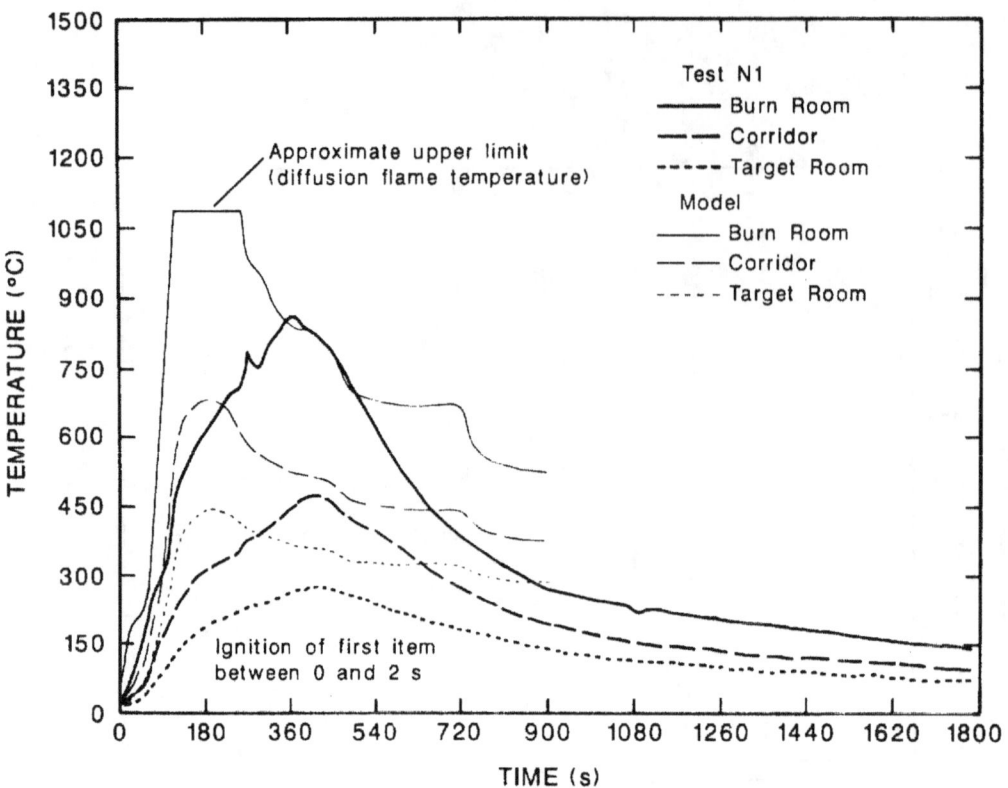

Figure 36. Model prediction and test result curves for the average upper layer temperature in the burn room/corridor/target room for NFR furnishings without the auxiliary burner.

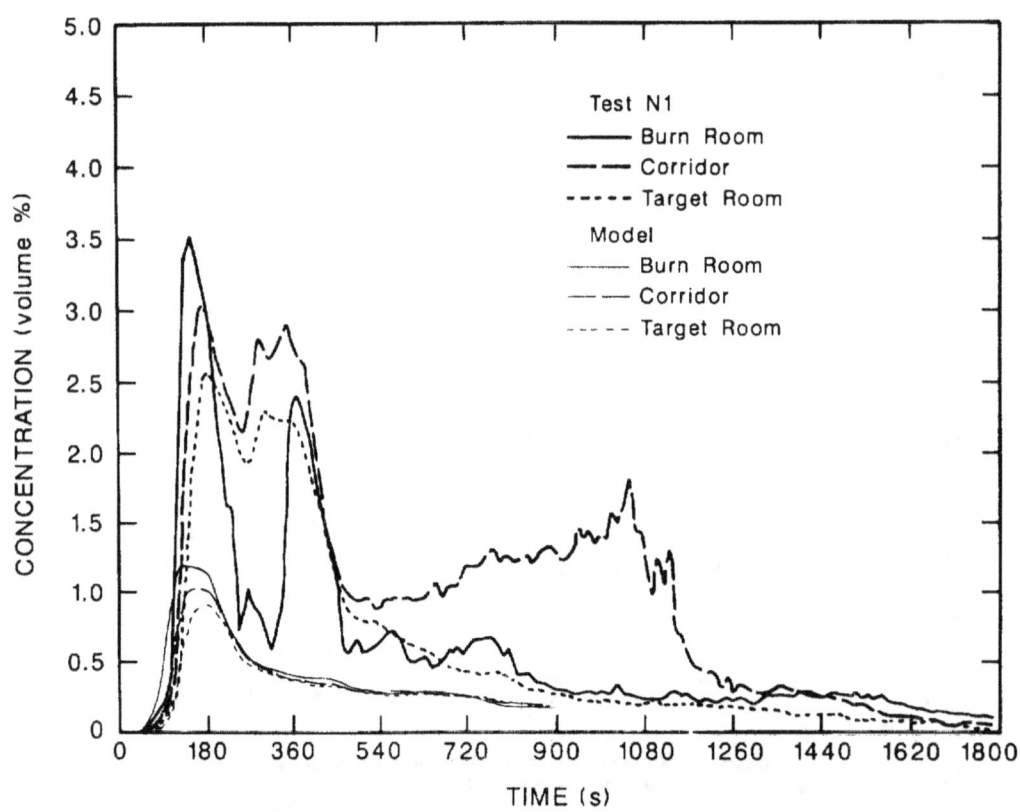

Figure 37. Model prediction and test result curves for carbon monoxide in the burn room/corridor/target room for NFR furnishings without the auxiliary burner.

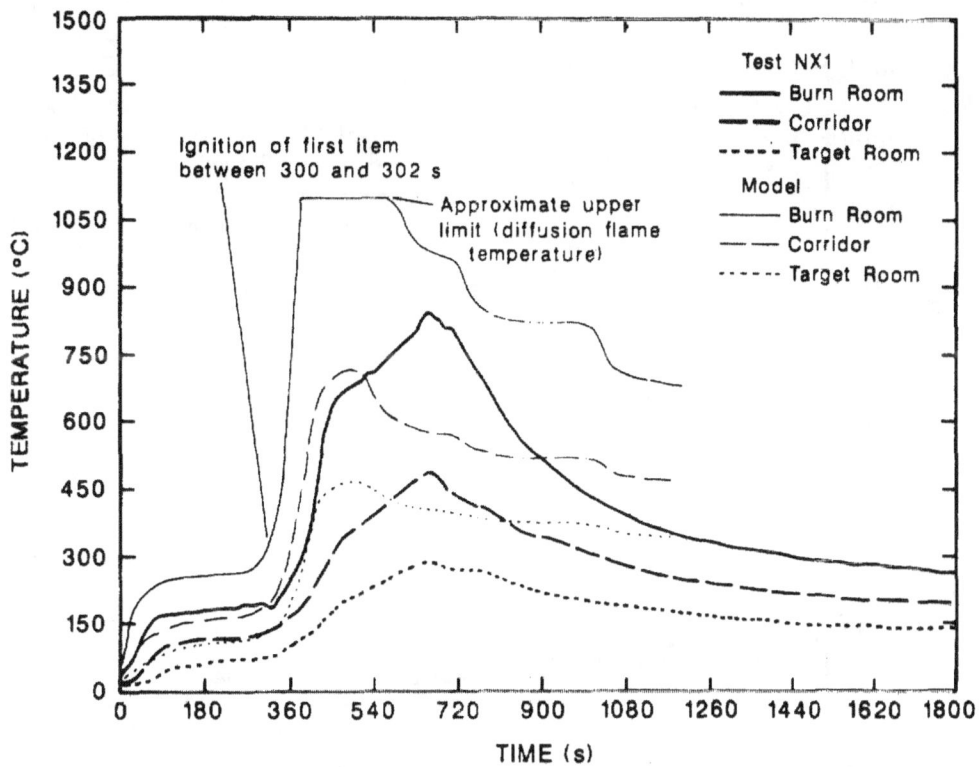

Figure 38. Model prediction and test result curves for the average upper layer temperature in the burn room/corridor/target room for NFR furnishings with the auxiliary burner.

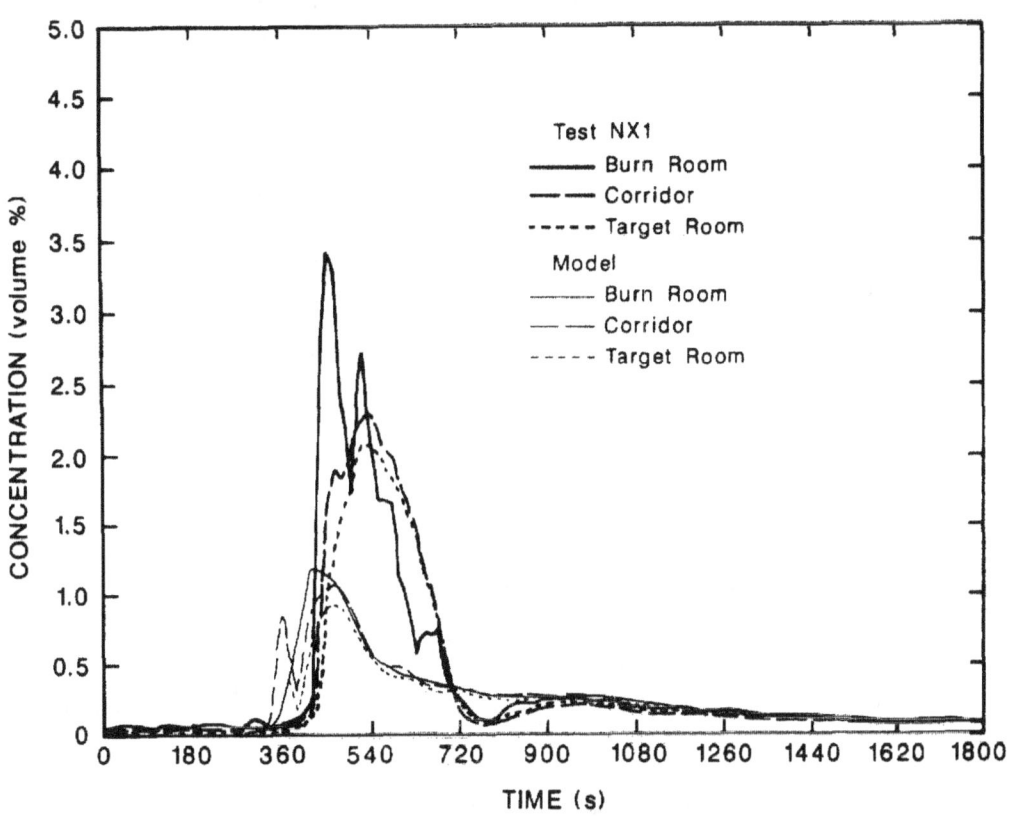

Figure 39. Model prediction and test result curves for carbon monoxide in the burn room/corridor/target room for NFR furnishings with the auxiliary burner.

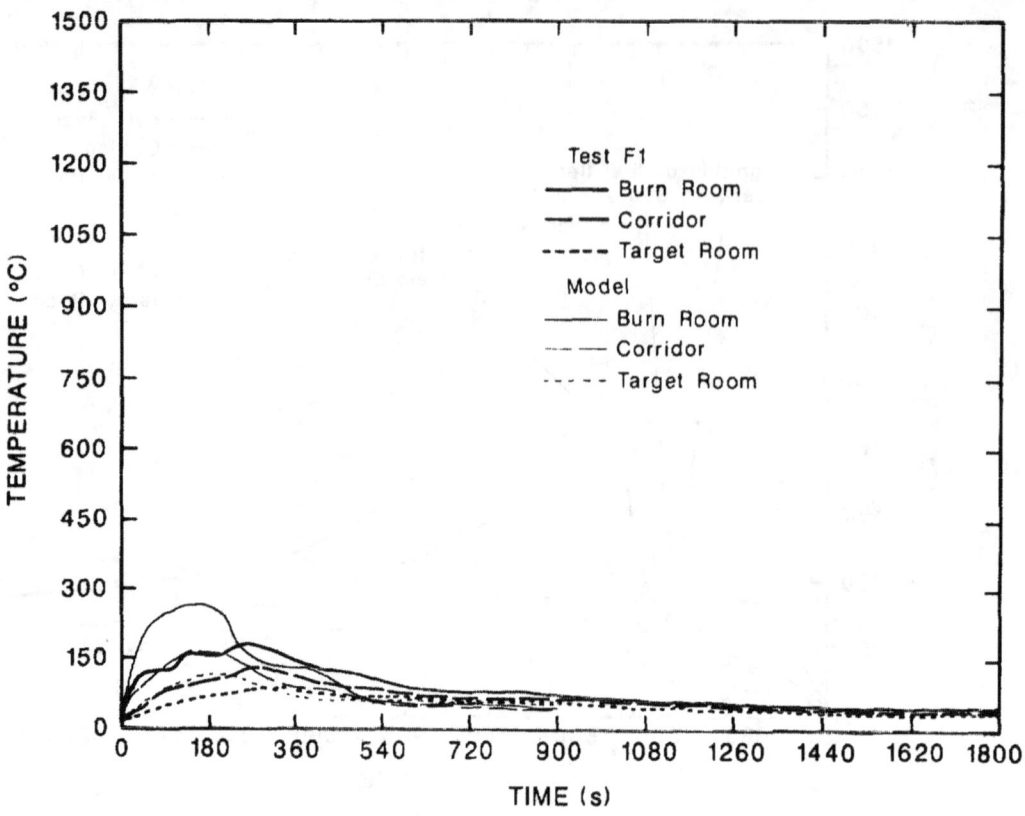

Figure 40. Model prediction and test result curves for average upper layer temperature in the burn room/corridor/target room for FR furnishings without the auxiliary burner.

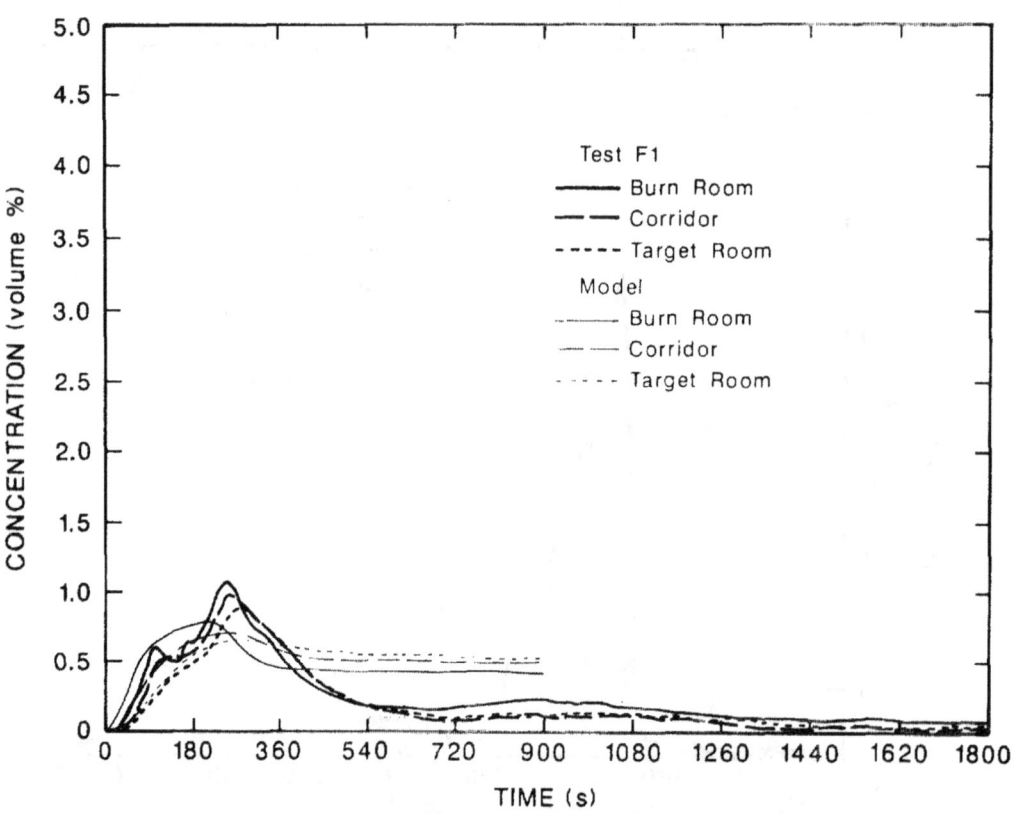

Figure 41. Model prediction and test result curves for carbon monoxide in the burn room/corridor/target room for FR furnishings without the auxiliary burner.

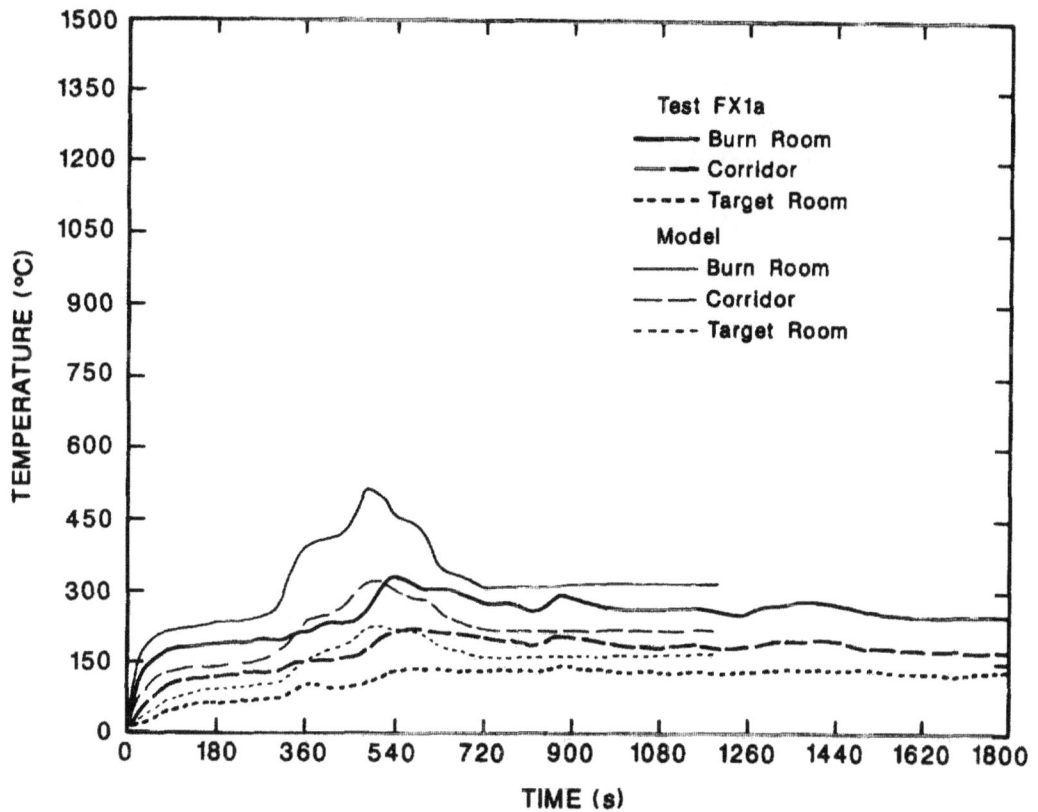

Figure 42. Model prediction and test result curves for the average upper layer temperature in the burn room/corridor/target room for FR furnishings with the auxiliary burner.

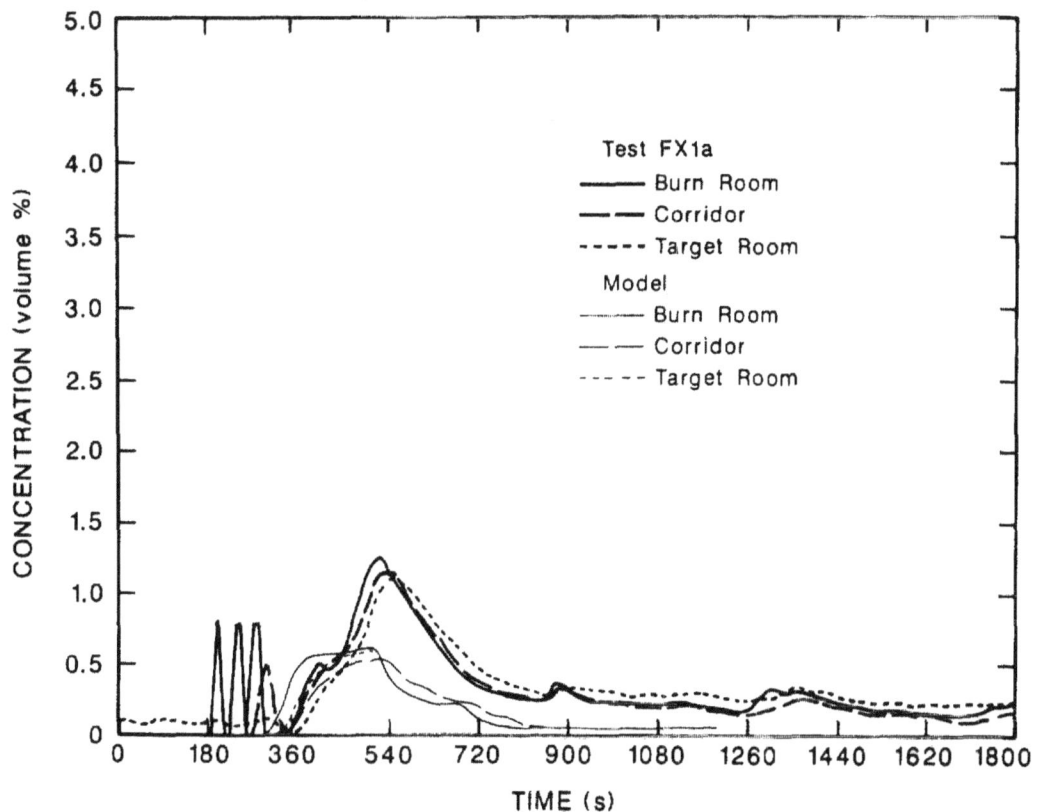

Figure 43. Model prediction and test result curves for carbon monoxide in the burn room/corridor/target room for FR furnishings with the auxiliary burner.

5

Large-Scale Tests

THE large-scale burn room/corridor/target room were designed to be the ultimate validation of the differences in hazard between the NFR and the FR components. An actual use environment might be rare where all five of the products studied would show up in one room. Nonetheless, it was felt that an arrangement where all the NFR or all the FR products are located in one test room would serve best to reveal the fire hazards expected. In either case, the fire was expected to be dominated by the more flammable products, with smaller, if any, contributions from the less flammable ones. Thus, the scenario is a challenging one, since it effectively exposes all the specimens to the fire conditions created by the most combustible one.

5.1 EXPERIMENTAL

5.1.1 ROOM/CORRIDOR/ROOM CONFIGURATION

The experimental facility is shown in Figure 29 and comprised a burn room connected to one end of a corridor and a target room at the opposite end of the corridor. A doorway from the target room led into a large open space over which a 3.7 m × 4.9 m hood, having an exhaust flow capacity of about 3 m^3/s, was used to collect the exhaust from the fire tests. Table 12 summarizes the dimensions of the facility. The ceilings and walls were constructed with gypsum board sheathing with a covering of 13 mm calcium silicate board to assure structural integrity during prolonged and repeated fire exposures. The concrete floors were covered with 13 mm gypsum board to protect the concrete. The construction materials used in this test series are given in Table 13 along with their properties.

5.1.2 ROOM FIRE TEST ARRANGEMENT

The arrangement of test items in the burn room is given in Figure 30. Each set of test items consisted of a chair, fabricated from four upholstered foam cushions; two television cabinets; two business machine cabinets; an array of twenty five power cables strung along a vertical steel ladder; and one circuit board panel mounted behind the other furnishings. This basic arrangement was the same for tests on both the NFR items and the FR ones. A 120 kW auxiliary natural gas burner in one back corner of the room simulated the combustion of additional furnishings in the room. A 50 kW natural gas burner, positioned under the television cabinets and in contact with one side of the chair, served as the ignition source.

Seven room fires were performed as follows, in the given order:

Test	Description
F1	FR items, with animals
FX0	FR items, with auxiliary burner, no animals
FX1	FR items, with auxiliary burner, with animals
N1	NFR items, with animals
FX1a	Repeat of FX1
NX0	NFR items, with auxiliary burner, no animals
NX1	NFR items, with auxiliary burner, with animals

For the above test numbers F = fire-retarded, N = non fire-retarded; X = with auxiliary burner, absence of X = no auxiliary burner; 0 = no animals, and 1 = with animals.

5.1.3 INSTRUMENTATION

The location of all instrumentation used in the test series are summarized in Table 14 with some of the instrumentation also shown in Figure 24. Data were recorded with an automatic data logging system at a rate of 120 data channels every 12 seconds.

5.1.3.1 Thermocouples and Pressure Probes

Six thermocouple trees, shown in Figure 29, were used in this study. These thermocouples were fabricated from 0.51 mm diameter Chromel and Alumel wires. Differential static pressure measurements were made on the wall surfaces between rooms near the floor. These pressure measurements between interconnecting rooms together with air temperatures measured by the interior and doorway thermocouple trees in the room with the outflow can be used to determine the neutral plane height and mass flow through the doorway [23]. The thermocouple trees also measured the thermal stratification in the room-corridor facility. In addition, a 0.51 mm Chromel-Alumel thermocouple was used on the ceiling in the burn room, directly over the test items, to further help estimate heat transfer to the room boundaries. Two 0.05 mm Chromel-Alumel thermocouples were also used in the NFR tests near the burn room ceiling to provide fast response temperature readings which were relatively free of radiation errors for a closer determination of room flashover time (defined here as 600°C air temperature at 0.10 m below the ceiling).

5.1.3.2 Flux Meters

Water-cooled flux meters of the Gardon type were used on the ceilings and floors of the burn room, corridor, and target room to help characterize the thermal flux environment in these spaces. In addition, Gardon type flux meters (pointing upward) were positioned at heights of 0.61 and 1.22 m above the floor in the burn room to estimate the fluxes, emitted from the hot air and heated surfaces in the upper portion of the room, that were received by the test items in the room.

Combustion Gases

Stainless steel lines having a 7.8 mm inside diameter were used to transport O_2, CO, and CO_2 in the burn room, corridor, and target room to gas analyzers for continuous measurement. Oxygen concentrations were monitored with paramagnetic analyzers, and CO and CO_2 were measured with infrared analyzers. In addition, combustion gases from the target room were routed through a 3.4 m long sampling line to three 200 l animal chambers where animals were exposed to the combustion gases and where analytical measurements of CO, CO_2, HCl, HCN, and HBr were taken. These chambers were located in a closed room adjacent to one side of the corridor. Another 2.8 m long sampling line to this room carried the combustion products from the burn room at a rate of 2.9 l/s for analytical measurements. The gases from both rooms were returned to the respective rooms. Both sampling lines were initially constructed of glass and polymethylmethacrylate tubing having a diameter of 51 mm. The first 2 m of the lines protruding into the rooms were glass, except for the last NFR run which required stainless steel tubing in the burn room to prevent melting. The animal chambers with their associated filling and exhaust lines and blower fan were equipment used in a previous study [4]. The filling times for these chambers in the initial three tests of the present study were found to be lower than desired. If it is assumed that a steady state flow of a combustion gas from the target room entered the animal chamber and was instantaneously mixed throughout the volume of the chamber prior to exiting at the opposite end of the chamber, then the chamber concentration of that gas can be estimated as a function of its concentration in the room and the filling rate and duration as follows:

$$C = \frac{\dot{m}}{\dot{V}_o}[1 - \exp(-\dot{V}_o t/V_c)] \qquad (5)$$

where:

C = combustion gas concentration (mg/l)
$\dot{m}$ = generation rate of combustion gas (mg/s)
t = chamber filling time (s)
$\dot{V}_o$ = volumetric airflow through chamber (l/s)
V_c = volume of chamber (l)

Inherent in the above equation is the assumption that there is no material loss due to surface deposition in the sampling line and chamber.

Calculated concentrations for a gas using the above equation, expressed in percent of its concentration in the target room, are given in Table 15 for filling times of 60, 120, and 180 s. These estimates confirmed that the filling rates for tests 1 to 3 were too low. Consequently, a higher capacity blower was used in the subsequent tests 4 to 6. Certain leaks in the chambers and in the blower line were also found prior to test 7. Resealing of the leaks resulted in different percentage fill in the chambers for test 7. Table 15 includes the estimated percent of the room concentration attained in the chambers for these tests at 60, 120, and 180 s.

5.1.3.4 Smoke

Smoke meters were not used within the burn room; instead, smoke performance was measured by determining the total amount of smoke collected through the exhaust hood system; see below.

5.1.3.5 Exhaust Hood Measurements

Temperatures, velocities, oxygen, CO, and CO_2 concentrations and smoke in the exhaust collection hood were monitored with the instrumentation listed in Table 14. These measurements were used to determine the total rate of heat production and to quantify the production of CO, CO_2 and smoke from combustion of the test items.

5.1.3.6 Gas Sampling Technique for Large-Scale Tests (Burn Room)

The same equipment was used for the large-scale tests as for the previously described Furniture Calorimeter tests. However, a modification to the impinger assembly was done to provide for continuous sampling during the burn. The apparatus is diagrammed in Figure 44. With the impingers connected in parallel, the gases were bubbled through one arm while the other arm was being replaced with a fresh impinger. The mass flow through the impingers was nominally 1 l/min and was verified prior to each test with a laboratory dry-gas meter. The ratio of gases collected to gases recirculated was nominally 1:173.

The impinger solution was 5 mM KOH for the first test and 10 mM KOH for the remaining tests. Sampling was initiated when the 50 kW natural gas burner was started and stopped when the fire was determined to be out. The samples were prepared and analyzed as previously described.

5.1.3.7 Gas Sampling Technique for Large-Scale Tests (Animal Exposure Boxes)

Gas samples from the three exposure boxes were collected at the nose levels of the animals. The apparatus is diagrammed in Figure 45. The flow was generated with a sampling pump and regulated with individual rotameters for each box. The flow rates (30-45 ml/min) were verified with a soap bubble meter before each test. Samples were collected for the duration of 10 to 30 minutes during the animal exposures.

For the first test one 3 ml microimpinger with 5 mM KOH was used. Because of a low pH value, the next test used two 3 ml microimpingers in series with 10 mM KOH. The third test used two micro impingers in series with 100 mM KOH. The remaining tests used two, tared 30 ml impingers in series containing approximately 25 ml of 10 mM KOH. The microimpingers were separated by a 25 mm PTFE filter with a 0.45 nominal porosity. The 30 ml impingers were separated with the same diameter PTFE filter with the same nominal porosity. At the end of the test the microimpinger solutions were quantitatively transferred and diluted to 5ml with 5ml KOH. The 30 ml impinger solutions were transferred to plastic containers. The filter with the collected soot and aerosols was added to the first impinger solution in all cases.

5.1.3.8 Animal Exposures

Three animal exposure chambers identical to those used in the NBS Toxicity Test Method were placed one on top of the other in a room located outside the room-corridor-room arrangement (Figure 29). A glass sampling line (51 mm in diameter and 3.4 m long) was located such that smoke was pulled from a position 0.34 m from the ceiling, 0.30 m from the east wall and 0.50 m from the south wall of the target room. The smoke was continuously transported through all three animal chambers with no animals present until the time designated for the animal exposures. Each chamber was equipped with gas analyzers for CO, CO_2, and O_2 as well as a thermocouple and an individual sampling port for HCN. A strip chart recorder was set up to continuously monitor the CO in one of the animal exposure chambers (the last to be closed) and this recording was used to determine the time that the sampling and return lines to each chamber were closed. The times that the chambers were closed and the estimated CO concentration at that time are given in Table 16. After the chamber was closed, the animals were inserted such that only their heads were exposed to the steady-state concentration for 30 minutes. In each experiment, three sets of six rats were exposed sequentially to different fractions of smoke drawn from the target room. The animals were watched during the exposures to determine the time of death and the survivors were examined following the 30-minute exposures for righting reflexes, eye reflexes, nose and mouth discharges, and respiratory effects. All surviving animals were kept and weighed daily for at least 14 days.

5.2 TEST PROCEDURE

In all of the tests, a 50 kW natural gas diffusion flame burner was used as the fire initiation source. A dry test meter was used to meter in the natural gas flow to the burner. The top side of the burner, which had a porous ceramic surface with nominal dimensions of 180 × 250 mm, was positioned 13 cm above the floor, against the side of the chair, and directly under the television cabinets as shown in Figure 25. In five tests, a 120 kW auxiliary diffusion flame burner was used in one back corner of the burn room. The top side of the burner, which had a 0.30 × 0.30 m porous ceramic surface, was placed 0.30 m above the floor. Both the fire initiation source and auxiliary burner used small pilot flames over their top surfaces for remote ignition.

Tests were performed with the data recording system turned on for about 120 s prior to ignition of the fire initiation source or of the auxiliary burner when it was used. In all of the tests, the 50 kW fire initiation source was left on for 200 s, and the fire was allowed to run for 1800 s. The auxiliary burner, when used, was turned on for 300 s prior to ignition of the 50 kW source and was left on for 2100 s.

5.3 LARGE-SCALE TEST RESULTS

5.3.1 DATA FROM EXHAUST STACK INSTRUMENTS

Figure 46 shows the basic rate of heat release data from the two tests without the auxiliary burner, N1 and F1. Figure 47 gives the corresponding data for the remaining tests, which all used the auxiliary burner.

The total mass flow rates of CO are given in Figures 48 and 49, while CO_2 flows are shown in Figures 50 and 51. Smoke production results are given in Figures 52 and 53.

5.3.2 DATA FROM ANIMAL CHAMBERS

In the total of seven large-scale tests, two (FXO and NXO) were performed without any animals to monitor the gas concentration and temperatures in the animal exposure chambers. The first three tests were with FR materials and the rates of the gas flow through the animal chambers were such that the highest levels of CO reached were approximately 3200 ppm. After these three tests, the pumps were changed so that the gas flow rates were higher and the concentration of CO in the animal chambers more closely resembled those in the target room.

The initial plan was to close the animal chambers when the CO concentrations were approximately 2000, 4000, and 6000 ppm. Any deaths at 2000 and 4000 ppm of CO would indicate the presence of additional toxic gases. Table 16 indicates the filling times and the CO levels estimated by the strip chart recorder in each animal chamber when it was closed. Table 17 shows the actual concentrations of CO and the other gases that were monitored.

In test F1, the chambers were closed at estimated CO concentrations of 3000, 5000, and 7000 ppm. However, the calibration of the CO strip chart recorder was incorrect and the chambers actually were closed with approximately half of the estimated CO concentrations. As seen in Table 17, the N-Gas Model prediction values based on the interaction of the concentration of monitored gases in the three sets of animal exposure chambers ranged from 0.15 to 0.66. If these gases were the only ones contributing to the toxicity of the atmospheres, no animals would be expected to die and, in fact, no lethalities were observed.

In test FXO, the strip chart CO recorder was corrected so that the animal chambers could be closed when the concentrations of CO were closer to the desired values. However, because of low animal chamber filling rates, the CO concentrations in the chambers leveled off before the maximum desired level could be reached. The animal exposure chambers were closed at such times that the measured average CO values in the chambers were 1600 ppm, and 2770 ppm, which was the highest attainable. The third chamber was then closed at a later time (almost 20 minutes following ignition) in order to try to obtain combustion products from some of the materials that would become involved later in the fire. Under these filling conditions, the N-Gas Model prediction values based on the monitored gases ranged from 0.51 to 0.78. Since this was a test without animals, the correctness of the N-Gas Model prediction could not be checked.

Test FX1 was a repeat of FXO except that animals were included. The exposure chambers were closed at times which resulted in average chamber CO levels of 1750, 2530, and 1060 ppm (the last chamber was closed at a much later time in the exposure to obtain some of the combustion products from the materials that became involved in the fire at that later time). The N-Gas Model prediction values for these three chambers ranged from 0.41 to 0.82; no animals were expected to die and none did.

Experiment FX1 was repeated and called FX1a. In this test, the pumps were changed so that the animal exposure chamber atmospheres more closely resembled that of the target room. In this experiment it was possible to close the animal exposure chambers at higher CO concentrations—average CO concentrations in the three animal exposure chambers were 2400, 4750, and 6000 ppm. The N-Gas Model prediction values were 0.70 with no deaths, 1.29 with 5/6 deaths, and 1.65 with 6/6 deaths. These results (deaths were observed only when the N-gas Model prediction value was above 1.0) indicate that no gases other than those monitored in these experiments were contributing to the toxic atmospheres (i.e., these atmospheres were not unusually toxic). The determination of extreme toxic potency (concentration of material decomposed which produces lethal conditions) was not possible under these experimental conditions since multiple materials were being burned, mass loss was not monitored, and knowledge of what materials were involved in the fire at what times was not known.

All of the NFR material tests were conducted with the higher gas flow rates. However in test N1, the pumps to the animal exposure chambers were not turned on until almost 3 minutes after ignition, at which point the chambers filled very rapidly. The bottom chamber was closed at 3.5 minutes after ignition, giving an average CO concentration of 7140 ppm. The middle chamber was closed at 4 min, giving 8880 ppm, while the top chamber was closed at 6.5 min and showed CO > 10,000 ppm. The average gas concentrations in the top chamber are not given in Table 17, since the animal exposure chamber leaked. In the middle chamber, the N-Gas Model prediction value was 4.10 or approximately four times the lethal concentrations. An interesting observation which needs more research to substantiate is that the animals died in approximately 1/4 the 30-minute exposure time. In other words, if the N-Gas Model prediction value was 1.0, one would expect half of the animals to die in the 30-minute exposure period. Perhaps if the N-Gas Model prediction value is 4.1, one could predict that the animals would die in (1/4.1) × 30 min or 7.3 min. In the middle chamber, the mean time to death was 6.5 ± 0.15 min. In the bottom chamber, the N-Gas Model prediction value was 3.16 or three times the lethal concentration of gases. As expected, all the animals died. If the time factor holds, the predicted time of death would have been 9.5 minutes. In actuality, the mean time to death was 10.2 ± 0.1 minutes. Since in the middle chamber, the animals died slightly earlier than expected and in the bottom chamber, they died slightly later than expected, it appears as though the gases generated in this exposure are probably responsible for the lethalities that occurred and there is no need to presume other gases are involved.

In experiment NXO, no animals were exposed. The bottom chamber was closed at 7 min 45 sec after ignition, giving an average CO concentration of 3400 ppm. The middle chamber was closed at 8:10, but was later found to leak. The top chamber was closed at 8:30, giving a concentration of 9150 ppm. The N-Gas Model prediction values for the bottom and top chambers were 3.11 and 3.62. These values would predict that if animals had been exposed, they would have died.

Experiment NX1 was the repeat of NXO with animals. In this experiment, the bottom and middle animal exposure chambers were closed at lower estimated CO values and the top chamber was not closed until later in the experiment to try to include the combustion products from the materials that would be involved during this later period of time. The actual average CO concentrations were 1820 ppm in the bottom chamber, 4490 ppm in the middle chamber and 1160 ppm in the top chamber. Since no animals died in the bottom chamber with an N-Gas Model prediction value of 0.97, there is no need to presuppose the involvement of gases other than those monitored. In the middle chamber, the N-Gas Model prediction value was 2.82 and all the animals died, as expected. If the above discussion on use of the N-Gas Model to predict time to death is valid, then the animals should have died 10.6 minutes into the exposure. The actual mean time to death was 9.3 ± 0.6 minutes which indicates that the deaths that occurred are probably attributable to the gases that were examined. The top chamber, although filled at the later time, did not produce any deaths at the N-Gas Model prediction value of 0.62, indicating that other gases could not be contributing to more than 38% of the toxicity. But since the bottom chamber had no deaths at an N-Gas value of 0.97, the likelihood of other gases being involved is low.

For all tests, a summary of the total integrated value of the gas exposure is provided in Table 18. The results are expressed as ppm·min.

5.3.3 DATA FROM OTHER OBSERVATIONS

The amount of specimen mass consumed was determined by examining the test residue after the test. These data are given in Table 19. Test F1, the FR test where the auxiliary burner was not used, resulted in only a very small amount (6.0 kg) of combustibles being consumed. Thus, as was expected in the test formulation, it did not achieve the goal of providing for substantial fire involvement to the furnishings. Consequently, the data from test F1 will not be used in assessing the performance.

5.4 COMPARISON BETWEEN SMALL-SCALE AND LARGE-SCALE TOXICITY FINDINGS

5.4.1 SPECIES YIELDS

Direct LC_{50} comparisons between the bench-scale and the large-scale findings would be the simplest basis for comparison. Unfortunately, due to the expense involved, complete bioassay studies, leading to an exact LC_{50} value, could be done in neither series of tests. The main question, however—do the test conditions in the bench scale accurately predict the evolution of toxic gases in the large-scale environment?—can be answered by some available data. Before this question is asked, it is appropriate to consider how well the less-than-room-scale measurements correlate amongst themselves. For this purpose, one can compare the species yield data for CO, CO_2, HCl, HBr, and HCN, as obtained in three devices:

- the Cone Calorimeter
- the Furniture Calorimeter
- the NBS combustion toxicity apparatus

Table 20 gives this comparison. Note that, in all cases, the measurements referred to are made *after* the tip of the flame. Thus, they represent the toxic effects of cooled combustion products and are not values in the flame zone (which would not be relevant to the study of toxic fire hazards).

The comparison for CO shows that, in those cases where the Cone Calorimeter data show very low values, the values obtained from the NBS combustion toxicity apparatus are typically higher and are closer to the values obtained in the Furniture Calorimeter. For the materials producing higher yields of CO, no systematic differences between data from the three devices are evident, although scatter is, in some cases, high.

For CO_2, the data from all three devices are generally in good agreement, with possibly a slightly better agreement obtained between the Cone Calorimeter and the NBS combustion toxicity apparatus than between the Furniture Calorimeter and the other two devices.

For the acid gases, the number of tests where such measurements needed to be taken were small, thus statistically valid conclusions are difficult to make. For HCN, based on three materials where data from more than 1 system were available, there is scatter of approximately a factor of 3. Part of the explanation lies in the small amount of HCN being produced. Another reason for differences in these two measurements could lie in the fact that different measurement techniques (gas chromatography and ion chromatography) were used. GC measurements will give only gaseous HCN yields. With the ion chromatography technique, however, a fraction of any HCN which is adsorbed onto soot particulates could be desorbed and included in the reported measurement.

For HCl, in the two cases where measurements from different systems were available, the results agreed very closely.

For HBr, the values recorded in the NBS combustion toxicity apparatus are substantially smaller than those obtained from either the Cone Calorimeter or the Furniture Calorimeter. Since there is no reason to presume that any greater losses would occur in the NBS combustion toxicity apparatus than in the other devices, the cause of this difference is not apparent. The hydrogen halide values even though showing good agreement for HCl and poor agreement for HBr, are not significant in either case from a toxicological point of view. That is, the contribution of these gases to the total toxicity was very slight for all the materials studied.

5.4.2 FRACTION OF TOXICITY ACCOUNTED BY CO

Another measure of the similarity or dissimilarity of the combustion environments associated with different tests is whether the fraction of the total toxicity accounted by CO is similar. If such ratios are similar for two tests, then it can be concluded that toxicity predictions based on data from one test are reasonably applicable to predicting the performance in the second test. The [CO toxicity]/[total toxicity] ratios were determined separately for the bench-scale NBS combustion toxicity apparatus and for the large-scale room-corridor-room tests, and are shown in Tables 21 and 22, respectively. Since both the bench-scale and the large-scale measurements, indicate that once CO, CO_2, O_2, HCN, HCl, and HBr are accounted for, significant additional toxic agents do not need to be invoked, the total toxicity figure can be derived from the measurements of these six gases. This is done on the basis of the equation given with Table 17, and includes the additional effect of CO_2 in potentiating CO action. By contrast, "CO toxicity" values are defined simply as:

$$\text{CO toxicity} = \frac{[\text{CO}]}{\text{LC}_{50}(\text{CO})}$$

For the NFR bench-scale tests, Table 21 shows that CO accounts for 17 to 44% of the total. In comparison, Table 22 shows that the large-scale NFR tests produced a range of 27 to 33%. For the FR products, the bench-scale range was 45 to 75%, while the large-scale range was 44 to 54%. The agreement is highly encouraging, and suggests that the effects of differing combustion conditions do not manifest themselves as preferential evolution of the major toxic agents in the different scale tests.

Even though CO *yields* are substantially underestimated by the NBS Combustion Toxicity Test (Table 26), it is not contradictory to observe that the fraction of toxicity accounted by CO is roughly similar. The differences in the two situations can be largely attributed to differences in CO_2 and O_2 levels not being the same. This happens since the dilution flows in the large-scale test are governed by both the geometry and the burning rates, and are not fixed as in a closed-box toxicity test method.

5.4.3 COMBUSTION PRODUCTS OF EXTREME TOXIC POTENCY

The presence, or absence, of extreme toxic potency was tested primarily using the NBS combustion toxicity test apparatus. None of the specimens, either NFR or FR were considered to show extreme toxic potency.

In the large-scale tests, a corresponding evaluation of the toxicity of the combustion products in units of mg/l could not be done since the environment was not a closed system. A reasonable, if somewhat indirect, assessment for the question of extreme toxicity is determined as follows. The results in the section below show that unusual toxic potency is not observed. Therefore, if extreme toxic potency were to occur, it would have to be because of the excessive amount of the known gases already being measured. The results in Tables 18 and 27, however, show that, compared to the NFR products, the FR products produced lower, not higher, quantities of the toxic gases monitored.

5.4.4 COMBUSTION PRODUCTS OF UNUSUAL TOXIC POTENCY

The animal data presented in Table 17 suggest that the 6-gas model is a reasonably complete representation of the toxic species encountered, and that unusual species are not active, at least in any major way. The data for NFR products show that at a 6-gas prediction value = 0.97 no animal deaths were observed; a close upper bound could not be obtained, but at 2.82 all animals died. For FR items, no mortalities were observed at a 6-gas prediction value = 0.70, while 56 deaths were observed at 1.29.

5.5 COMPARISON BETWEEN COMPUTER MODEL PREDICTIONS AND ACTUAL LARGE-SCALE RESULTS

Some comparisons between the computer predictions and the experimental findings for the NFR and FR products are also given in Figures 36 to 43. With regard to the average upper layer temperature for the FR furnishings without the auxiliary burner (Figure 40), the chosen model scenario predicts higher temperatures in the burn room than were actually achieved in the large-scale tests. In this test (F1) the model predicted that the upper layer temperature would reach almost 300°C, but the actual temperature reached was only about 170°C. In all of the large-scale tests, with or without the auxiliary burner or using NFR or FR furnishings, the same result is found. This result is true not only for the burn room, but also for the corridor and target room.

Like all of the current zone-based fire models, including the FAST model [21], the over prediction of upper layer temperature is not surprising. Upper layer temperatures are typically over-predicted by these models. Too hot an upper layer may be caused either by over-prediction of the heat entering the layer from the fire plume, or by under-prediction of the heat lost from the layer by radiation, conduction, or convection. In FAST, like other contemporary models, mixing occurs only at the vents between rooms and no heat is lost by radiation to the lower layer. While the temperature of the lower layer may be important when considering the effect on occupants and equipment, currently there cannot be a good prediction, due to the assumption in the model of no radiative heating and no mixing into this zone (except at the vents). Thus, the lower layer temperature is correspondingly under-predicted by the model. As the models become more sophisticated, these simplifying assumptions are being removed. In a newer version of FAST, the lower layer can be heated via radiation and mixing. Improved comparability of layer temperatures in another series of tests was noted by Jones and Peacock [24].

In regard to the carbon monoxide content in the rooms, the model scenario predicts less than is actually produced. For instance, the curves in Figure 41 for FR furnishings without the auxiliary burner indicate a predicted CO concentration in the burn room of ca. 0.7 volume % whereas the actual CO content was ca. 1.1 volume %. In the case of the NFR furnishings, the actual CO values are more than twice as great as the model predicts, as can be seen in Figures 37 and 39. Since all the inputs to the model on burning rates and CO production for the individual burning items came from free burn data, the effects of vitiated burning are not taken into account. Without sufficient oxygen for complete burning, higher CO concentrations are expected.

As an overall assessment of the usefulness of the model predictions, several key points are obvious. First, the modeling exercise reported herein represents one of the first uses of predictive fire modeling in the design of large-scale tests. The choice of the final design for the room/corridor/room test facility was greatly simplified by the use of the model, eliminating alternative design options without costly construction and testing. Secondly, the choice of the input parameters for the model is critical in providing meaningful model outputs. Significant care must be taken in the use of free burn data for multiple burning items to accurately model the ignition and burning of more than one item in an enclosure. Finally, the model user must understand the limitations and simplifying assumptions underlying the design of the predictive model to be able to assess the meaning of the outputs from the model.

TABLE 12 Dimensions of Room/Corridor/Room Large-Scale Test Facility.

Location	Dimensions (m)
Burn room	2.44 W × 3.69 L × 2.44 H
Burn room doorway	0.76 W × 2.01 H
Corridor	2.44 W × 3.69 L × 2.44 H
Corridor doorway	0.76 W × 2.02 H
Target room	2.44 W × 4.60 L × 2.44 H
Target room doorway	0.76 W × 2.03 H

TABLE 13 Construction of Large-Scale Test Facility.

Location	Material	Thickness t (mm)	Density ϱ (kg/m^3)	Heat Cap. C_p (kJ/kg-°C)	Thermal Cond. k (W/m-°C)
Ceiling and Walls	calcium silicate board	12.7	720	1.3	0.12
Floor	gypsum board	12.7	930	1.1	0.17

TABLE 14 Location of Instrumentation.

I. Room-Corridor

  A. Thermocouples

   Tree 1 in burn room—ten 0.51 mm thermocouples at 0, 0.15, 0.60, 0.91, 1.23, 1.52, 1.83, 2.13, 2.30, and 2.44 m above floor

   Tree 2 in burn room doorway—ten 0.51 mm thermocouples at 0.15, 0.30, 0.61, 0.76, 0.91, 1.07, 1.22, 1.37, 1.52, and 1.83 m above floor.

   Tree 3 in corridor—twelve 0.51 mm thermocouples at 0, 0.15, 0.66, 0.97, 1.11, 1.23, 1.42, 1.57, 1.89, 2.03, 2.15, and 2.44 m above floor.

   Tree 4 in corridor doorway—ten 0.51 mm thermocouples at 0.15, 0.30, 0.61, 0.76, 0.91, 1.07, 1.22, 1.37, 1.52, and 1.83 above floor.

   Tree 5 in target room—eleven 0.51 mm thermocouples at 0, 0.15, 0.60, 0.91, 1.06, 1.23, 1.52, 1.83, 2.13, 2.30, and 2.44 m above floor.

   Tree 6 in target room doorway—eight shielded thermocouples at 0.15, 0.61, 0.91, 1.07, 1.22, 1.52, 1.83, and 1.93 m above floor.

   Two 0.05 mm thermocouples, near tree 1 and 0.10 m down from ceiling in burn room.

   One 0.51 mm thermocouple on middle of ceiling in burn room.

   One thermocouple in top animal chamber.
   One thermocouple in middle animal chamber.
   One thermocouple in bottom animal chamber.

  B. Static pressure probes

   Pressure probes were placed, 0.08 m above the floor, along the wall between the burn room and corridor, between the corridor and target room, and between the target room and the outside.

(continued)

TABLE 14 (continued).

C. Flux meters

Burn room—three flux meters, 0.61 m from back wall at 0, 0.61, and 1.22 m heights above floor. One flux meter, on floor, 0.61 m from front wall. One flux meter at the center of the ceiling.

Corridor—one flux meter at center of floor. One flux meter at center of ceiling.

Target room—one flux meter at center of floor. One flux meter at center of ceiling.

D. Gas probes

Burn room—stainless steel probe (6.4 mm inside diameter) for CO, CO_2, and O_2 at 0.30 m from ceiling, 0.30 m from front wall and 0.48 m from side wall.

Burn room—glass line (51 mm inside diameter and 2.8 m long) for sampling of CO, CO_2, HBr, HCl, HCN at 0.30 m from ceiling, 0.30 m from front wall and 0.38 m from side wall.

Corridor—stainless steel probe (6.4 mm diameter) for CO, CO_2, and O_2 at 0.30 m from ceiling, 0.28 m from back wall and 0.42 m from side wall.

Target room—stainless steel probe (6.4 mm diameter) for CO, CO_2, and O_2 at 0.30 m from ceiling, 0.30 m from back wall and 0.29 m from side wall.

Target room—glass line (51 mm diameter and 3.4 m long) for sampling CO, CO_2, HBr, HCl, HCN at 0.34 m from ceiling, 0.30 m from back wall and 0.50 m from side wall.

II. Exhaust Hood

1—smoke meter
1—probe for sampling CO, CO_2, and O_2
9—pitot-static probe
9—thermocouples

TABLE 15 Ratio of Gases in Animal Chambers to Target Room for Specified Times.

Test 1 to 3					
				conc. in box/ conc. in room (%)	
Chamber	V(l)	$\dot{V}°$(l/s)	60 s	120 s	180 s
Top	200	0.411	12	22	31
Middle	200	0.411	12	22	31
Bottom	200	0.411	12	22	31
Test 4 to 6					
				conc. in box/ conc. in room (%)	
Chamber	V(l)	$\dot{V}°$(l/s)	60 s	120 s	180 s
Top	200	1.19	30	51	66
Middle	200	3.91	69	90	97
Bottom	200	3.91	69	90	97
Test 7					
				conc. in box/ conc. in room (%)	
Chamber	V(l)	$\dot{V}°$(l/s)	60 s	120 s	180 s
Top	200	2.88	58	82	93
Middle	200	4.22	72	92	98
Bottom	200	2.88	58	82	93

TABLE 16 Animal Exposure Chamber Filling Times and Estimated CO When Closed.

Test No.	Fire Retard.	Auxiliary Burner	Animal Chamber	Filling Times (min:sec)	Estimated CO[a] (ppm)
N1	—	—	Bottom	2:50– 3:30	3600
			Middle	2:50– 4:00	6800
			Top	2:50– 6:30	>10000*
NXO	—	+	Bottom	0– 7:45	1100
			Middle	0– 8:10	3400
			Top	0– 8:30	6000
NX1	—	+	Bottom	0– 7:35	1000
			Middle	0– 7:55	4000
			Top	0–15:00	1500
F1	+	—	Top	0– 3:30	3000[b]
			Middle	0– 4:45	5000[b]
			Bottom	0– 5:55	7000[b]
FXO	+	+	Top	0– 5:00	2500
			Middle	0– 8:45	3200*
			Bottom	0–19:45	2100
FX1	+	+	Top	0– 9:30	2000
			Middle	0–13:00	3200*
			Bottom	0–41:00	1500
FX1a	+	+	Bottom	0– 8:00	2000
			Middle	0– 9:05	4000
			Top	0– 9:55	6500

[a]CO concentrations on strip chart recorder used to determine time of animal chamber closure.
[b]Strip chart calibrated incorrectly such that CO concentration was actually half of that estimated from strip chart.
*Maximum level of CO in animal chambers.

TABLE 17 Chemical and Toxicological Results in Animal Exposure Chambers Filled During Large-Scale Room Burns.

Test No.	Fire Retard.	Auxiliary Burner	Animals Present	Animal Chamber	chamber Temp. (°C)	Gas Concentrations[a]						Deaths		6-Gas Prediction Within Exp.
						CO (ppm)	CO_2 (ppm)	HCN (ppm)	O_2 (%)	HBr (ppm)	HCl (ppm)	Within Exp.	Within & Post-Exp.	
N1	—	—	+	Top	28	—[b]	—[b]	78[c]	—[b]	ND	92	6/6	6/6	—
			+	Middle	24	8880	55400	240	13.9	—	—	6/6	6/6	4.10
			+	Bottom	22	7140	39300	200	16.1	ND	ND	6/6	6/6	3.16
NX0	—	+	—	Top	NM	9150	64500	160	12.3	ND	170	—	—	3.62
			—	Middle	23	—[b]	—[b]	—[b]	—[b]	—[b]	—[b]	—	—	—
			—	Bottom	22	5470	49900	185	14.0	ND	400	—	—	3.11
NX1	—	+	+	Top	31	1160	44100	>1	15.6	ND	ND	0/6	0/6	0.62
			+	Middle	23	4490	52700	200	13.9	ND	ND	6/6	6/6	2.82
			+	Bottom	20	1820	31600	50	16.6	ND	11	0/6	0/6	0.97
F1	+	—	+	Top	NM	390	2200	8	20.5	24	Trace	0/6	0/6	0.15
			+	Middle	NM	2260	9200	23	19.0	44	ND	0/6	0/6	0.66
			+	Bottom	NM	2500	8350	22	19.6	31	ND	0/6	0/6	0.64
FX0	+	+	—	Top	23	1610	13100	11	18.8	60	17	—	—	0.51
			—	Middle	23	2770	17100	16	18.1	22	4	—	—	0.78
			—	Bottom	22	1530	14700	18	18.6	18	15	—	—	0.54
FX1	+	+	+	Top	23	1750	16200	11	18.3	27	ND	0/6	0/6	0.56
			+	Middle	23	2530	21000	18	17.2	18	2	0/6	0/6	0.82
			+	Bottom	23	1060	12600	15	19.0	4	20	0/6	0/6	0.41
FX1a	+	+	+	Top	23	6000	27800	30	16.7	43	ND	6/6	6/6	1.65
			+	Middle	23	4750	21600	24	17.0	46	ND	5/6	5/6	1.29
			+	Bottom	21	2400	16800	12	18.1	54	ND	0/6	0/6	0.70

[a]Average concentration over 30-minute exposure.
[b]Animal exposure chamber leaked.
ND Not detected.
NM Not measured.
N-Gas prediction based on equation:

$$\frac{m[CO]}{[CO_2]-b} + \frac{[HCN]}{160} + \frac{21-[O_2]}{21-5.4} + \frac{[HBr]}{3000} + \frac{[HCl]}{3700}$$

where m = slope of LC_{50} line of CO in the presence of CO_2 and b = the y intercept of this line. Values used for m (−18.4 or 22.7 when CO_2 is below or above 5%, respectively) and b (122,000 or −39,000 when CO_2 is below or above 5%, respectively) were based on data obtained after 11986. 160 ppm, 5.4%, 3000 ppm and 3700 ppm are the 30-minute LC_{50} values of HCN, O_2, HBr, and HCl, respectively.

TABLE 18 Comparison of Peak Times and Concentration Areas for CO, HCl, HBr and HCN in Large-Scale Tests (Burn Room).

Test Number	CO		HCl		HBr		HCN	
	Peak[a] Time (min)	Area (ppm·min)	Peak[a] Time (min)	Area (ppm·min)	Peak[a] Time (min)	Area (ppm·min)	Peak[a] Time (min)	Area (ppm·min)
N1	2.8	146980	1.5	3700	—	—	1.5	2670
NX0	2.5	152220	3	5960	—	—	3	2920
NX1	2.7	114800	7	8930	—	—	3	2650
F1	4.5	74870	12.5	860	2.5	4160	2.5	800
FX0	3.3	118520	35	7230	4	4220	14	1170
FX1	2.8	109820	9	5820	9	4290	7.5	1200
FX1a	3.8	112200	26	14880	4	5710	4	2100

[a]Measured from time of ignition.

TABLE 19 Specimen Mass Consumed in Large-Scale Tests.

Test No.	TV (H/G)	Bus. mach. (F/A)	Chair (T/S)	Cables (D/K)	Ckt. Bd. (C/L)	Total (kg)	Avg. (kg)
N1	3.6	3.2	5.2	4.6	13.4	30.0	
NX0	3.6	3.2	5.2	4.6	13.4	30.0	30.0
NX1	3.6	3.2	5.2	4.6	13.4	30.0	
F1	2.1	0	3.9[a]	0	0	6.0	
FX0	2.1	0	9.5[a]	0	0	11.6	
FX1	2.1	0	9.5[a]	0	0	11.6	12.0
FX1a	2.1	0	10.8[a]	0	0	12.9	

[a] Estimated from observations of video at 1800 s for F1 and 2100 s for remaining tests

TABLE 20 Comparison Between Yields of Toxic Species in the Different Devices.

Specimen	NFR /FR	CO (kg/kg) Cone Cal.	CO Furn. Cal.	CO Tox. Test.	CO_2 (kg/kg) Cone Cal.	CO_2 Furn. Cal.	CO_2 Tox.[d] Test.	HCN (kg/kg) Cone Cal.	HCN Furn. Cal.	HCN Tox. Test.	HBr (kg/kg) Cone Cal.	HBr Furn. Cal.	HBr Tox. Test.	HCl (kg/kg) Cone Cal.	HCl Furn. Cal.	HCl Tox. Test.
TV Cabinet H	NFR	0.015	0.12	0.084	2.28	1.39	2.09	—	—	—	—	—	—	—	—	—
TV Cabinet G	FR	0.109	0.37	0.18	0.67	0.74	0.78	—	—	—	0.069	0.082	0.017	—	—	—
Bus. Machine F	NFR	0.037	0.13	0.17	2.21	1.61	1.98	—	—	—	—	—	—	—	—	—
Bus. Machine A	FR	0.055	0.29	0.30	1.60	1.45	1.53	—	—	—	—	—	—	—	—	—
Chair T	NFR	0.020	0.01	—	1.62	1.89	—	0.002	0.001	—	—	—	—	—	—	—
Chair S	FR	0.051	[a]	—	0.964	[a]	—	0.005	—	—	—	—	—	0.023	—	—
Chair T[b]	NFR	0.016	—	0.025	1.71	—	2.05	0.002	—	0.0007	—	—	—	—	—	—
Chair S[b]	FR	0.055	—	0.15	0.81	—	1.19	0.0023	—	0.0032	—	—	—	0.022	—	—
Cable D	NFR	0.041	0.12	—	1.77	1.61	—	—	—	—	—	—	—	0.112	0.121	—
Cable K	FR	0.060	0.10	—	1.34	1.04	—	—	—	—	—	—	—	0.131	0.133	—
Cable D[c]	NFR	0.029	—	0.050[e]	2.19	—	2.38	—	—	—	—	—	—	ND	—	—
Cable K[c]	FR	0.135	—	0.13	1.00	—	1.26	—	—	—	—	—	—	0.093	—	—
Circuit Bd. C	NFR	0.014	0.10	0.075[e]	2.07	1.71	2.13	—	—	—	—	—	—	—	—	—
Circuit Bd. L	FR	0.103	0.10	0.15	0.87	1.36	1.24	—	—	—	0.022	—	0.0043	—	—	—

[a] Could not be determined reliably.
[b] Foam only, no cover fabric.
[c] Wire insulation only.
[d] Determined only from those tests where animals were not used.
[e] Excludes data from the highest mass loading tested, since presumed unrepresentative.
— Not run
ND Not detected

TABLE 21 Fraction of Total Toxicity Accounted for by CO (NBS Combustion Toxicity Test).

Sample	FR	CO toxicity/total toxicity
TV Cabinet H	NFR	0.42
TV Cabinet G	FR	0.75
Bus. Machine F	NFR	0.53
Bus. Machine A	FR	0.66
Chair T (foam)	NFR	0.17
Chair S (foam)	FR	0.45
Cable D (wire ins.)	NFR	0.35
Cable K (wire ins.)	FR	0.62
Circuit Bd. C	NFR	0.44
Circuit Bd. L	FR	0.68

All data represent averages of all replicates conducted in bench scale.

TABLE 22 Fraction of Total Toxicity Accounted for by CO (Large-Scale).

Test	Animal chamber	CO toxicity/total toxicity
N1	top	0.27
	middle	0.33
	bottom	0.34
		avg. = 0.31
NX0	top	0.38
	middle	N.A.
	bottom	0.27
		avg. = 0.33
NX1	top	0.28
	middle	0.24
	bottom	0.28
		avg. = 0.27
F1	top	0.39
	middle	0.52
	bottom	0.59
		avg. = 0.50
FX0	top	0.48
	middle	0.54
	bottom	0.43
		avg. = 0.48
FX1	top	0.47
	middle	0.47
	bottom	0.39
		avg. = 0.44
FX1a	top	0.55
	middle	0.56
	bottom	0.52
		avg. = 0.54

TABLE 23 Times to Reach Untenable Conditions in Large-Scale Tests.

Test Number	Burn room Flashover[a] (s)	Burn room CO FED (s)	Target room CO FED (s)
N1	110	164	200
NX0	112	167	215
NX1	116	168	226
F1	∞ (185)[c]	∞ (0.41)[d]	∞ (0.33)[d]
FX0	∞ (273)[c]	1939	∞ (0.40)[d]
FX1	∞ (285)[c]	2288	∞ (0.29)[d]
FX1a	∞ (334)[c]	1140	1013

[a]Time when temperature reached 600°C
[b]Auxiliary burner output exceeds this flux
[c]Maximum burn room temperature (°C)
[d]Maximum CO FED attained
FED Fractional Effective Exposure-dose

TABLE 24 Comparison of Total Heat Release from Large-Scale Fires with Furniture Calorimeter and Cone Calorimeter Calculated Values.

Test Number	Total Heat Release (MJ) Large-Scale[a] Individual	Large-Scale[a] Average	Furn.	Cone
N1	639			
NX0	479	542	730	752
NX1	507			
F1	69		—[b]	94[c]
FX0	141			
FX1	116	121	—[b]	199[c]
FX1a	105			

[a]Corrected for auxiliary burner (252 MJ) and igniting torch (10 MJ)
[b]TV cabinet and chair only two items involved in fire; since Δh_c for chair could not be determined from Furniture Calorimeter tests, the result is indeterminate.
[c]Computed from TV cabinet and portion of chair consumed at 1800 s for F1 and 2100 s for FX0, FX1, and FX1a.

TABLE 25 Comparison of Smoke from Large-Scale Fires with Furniture Calorimeter and Cone Calorimeter Calculated Values.

Test	Large-scale Smoke prod. (m²)	Avg.	Large-scale Smoke yield (m²/kg)	Avg.	Furn. Cal. Smoke yield (m²/kg)	Cone Cal. Smoke yield (m²/kg)
N1	10540		351			
NX0	8795	9900	293	330	486	780
NX1	10360		345			
F1	7148		1191		1097	970
FX0	12630		1089			
FX1	12800	12400	1103	1038	638	725
FX1a	11890		922			

TABLE 26 Comparison of Average CO from Large-Scale Fires with Furniture Calorimeter, Cone Calorimeter, and Toxicity Test Calculated Values.

Test Number	Large-scale				Furn. Cal. CO yield (kg/kg)	Cone Cal. CO yield (kg/kg)	Tox. Test CO yield (kg/kg)
	CO prod. (kg)	Avg.	CO yield (kg/kg)	Avg.			
N1	6.6		0.22				
NX0	5.5	5.5	0.18	0.18	0.09[a]	0.02	0.074
NX1	4.3		0.14				
F1	1.3		0.22		—[b]	0.07	0.046
FX0	2.6		0.23				
FX1	2.7	2.8	0.23	0.23	—[b]	0.06	0.155
FX1a	3.0		0.23				

[a]Based on high values from Furniture Calorimeter which cannot be explained.
[b]TV cabinet and chair only two items involved in fire; since CO for chair could not be determined from Furniture Calorimeter tests, the result is indeterminate.

TABLE 27 Total Toxicity Results in Large-Scale Tests.

Test No.	CO prod. (kg)	CO toxicity/total toxicity	total toxicity (CO-equiv. kg)	avg. tot. tox. (CO-equiv. kg)
N1	6.6	0.31	21	
NX0	5.5	0.33	17	18
NX1	4.3	0.27	16	
F1	1.3	0.50	2.6	
FX0	2.6	0.48	5.5	
FX1	2.7	0.44	6.1	5.7
FX1a	3.0	0.54	5.6	

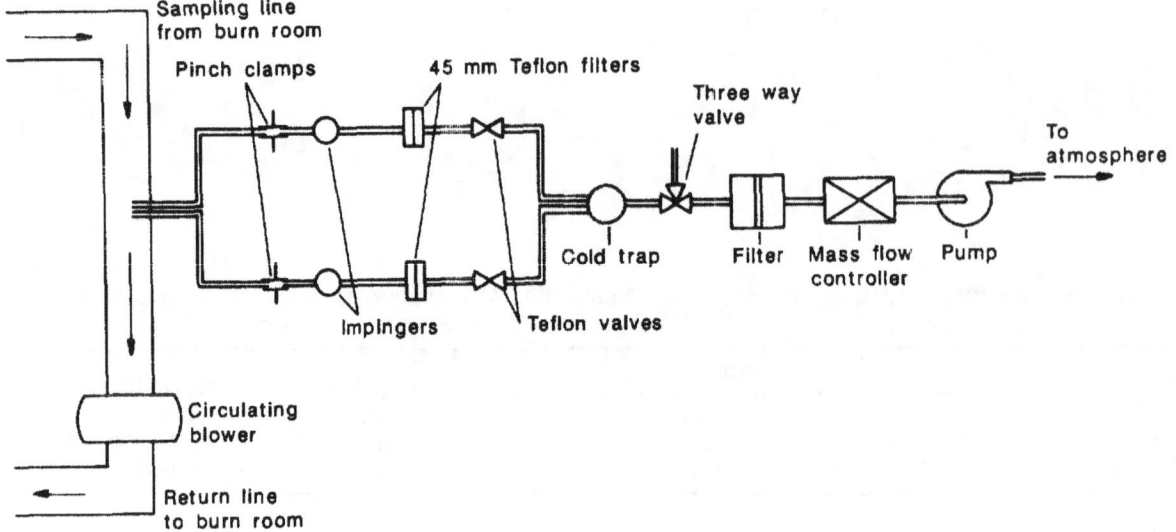

Figure 44. Impinger gas sampling in the large-scale burn room.

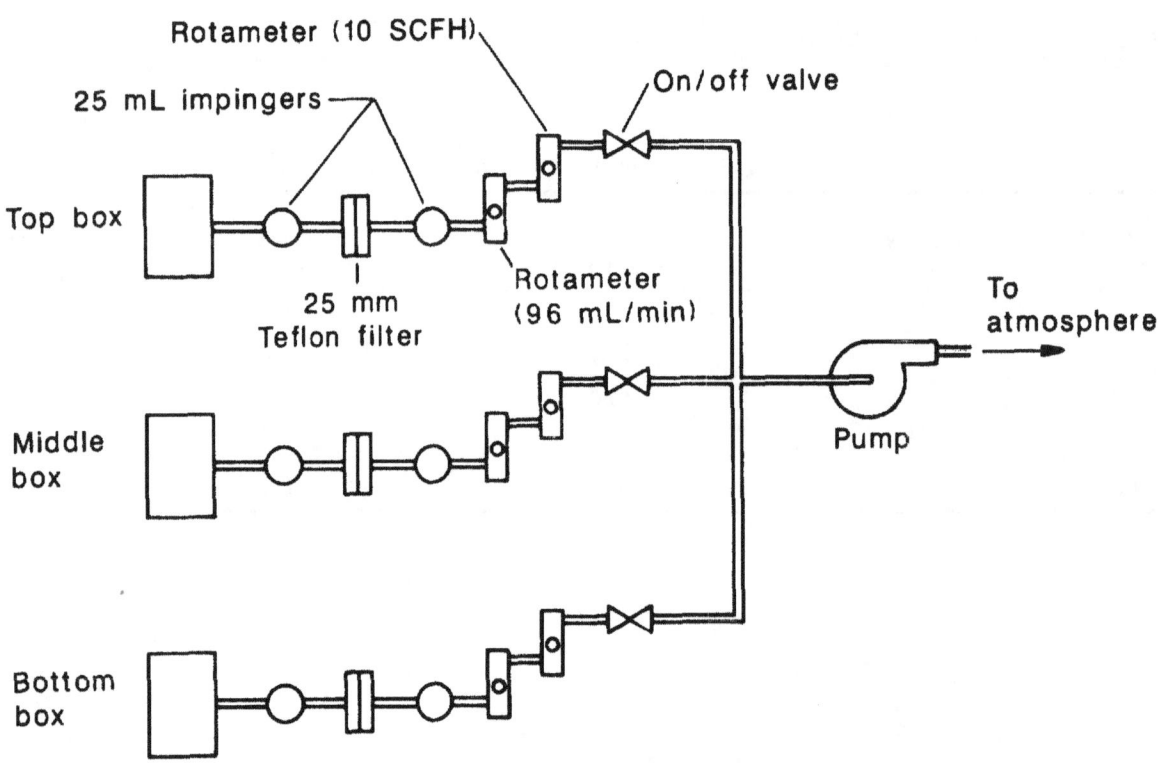

Figure 45. Impinger gas sampling in the large-scale animal chamber.

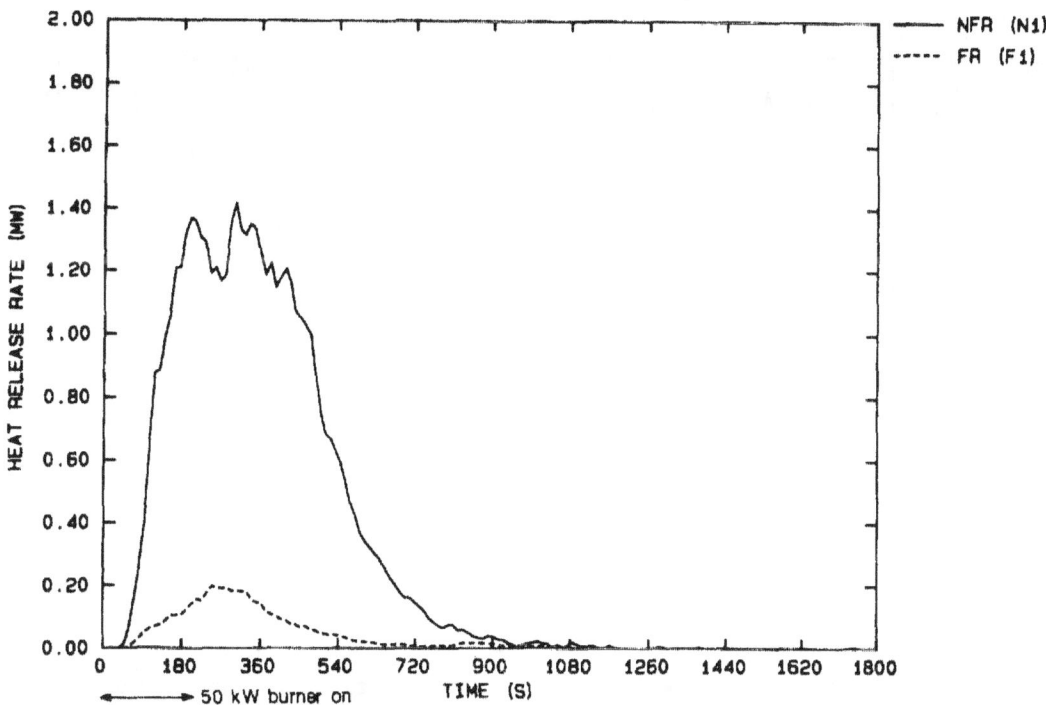

Figure 46. Heat release rates in the large-scale room/corridor/room tests without the auxiliary burner.

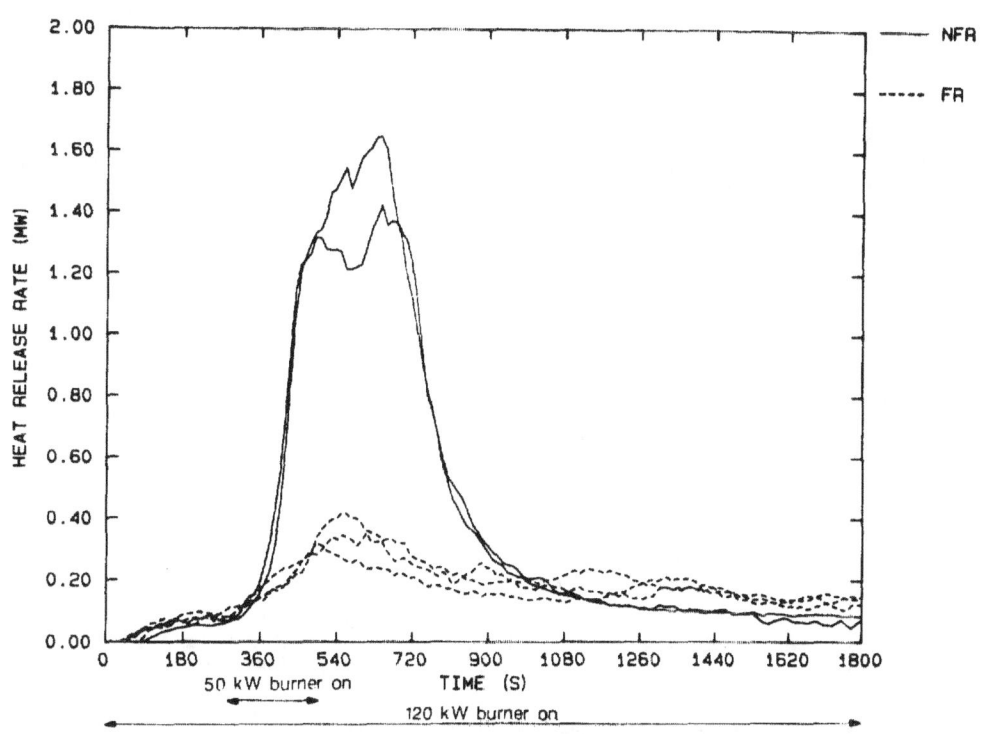

Figure 47. Heat release rates in the large-scale room/corridor/room tests with the auxiliary burner.

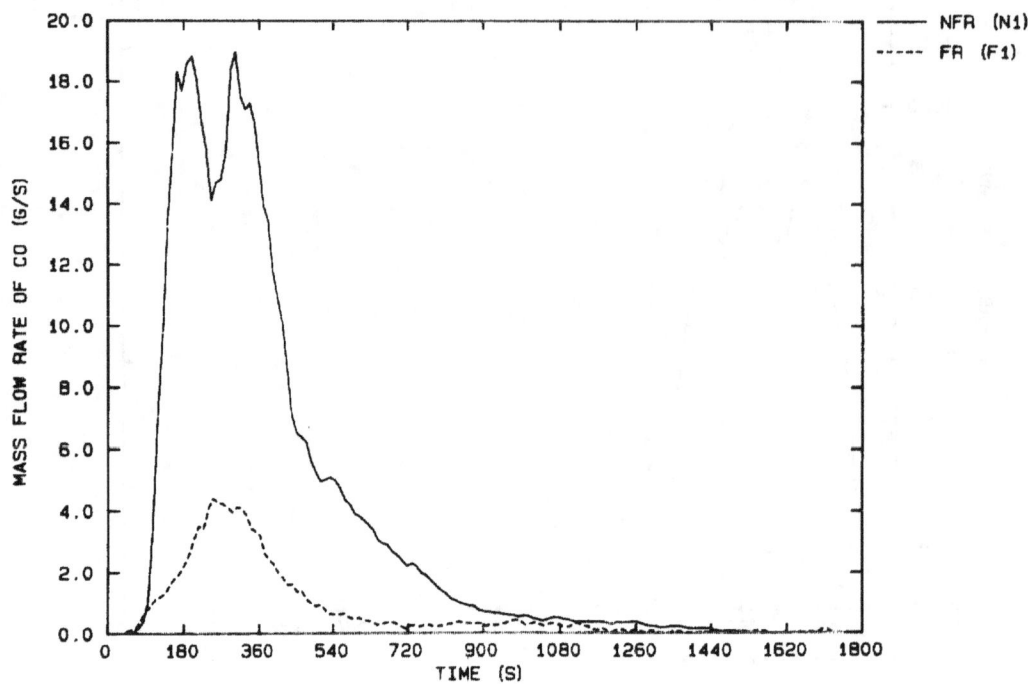

Figure 48. Mass flow rates of CO in the large-scale room/corridor/room tests without the auxiliary burner.

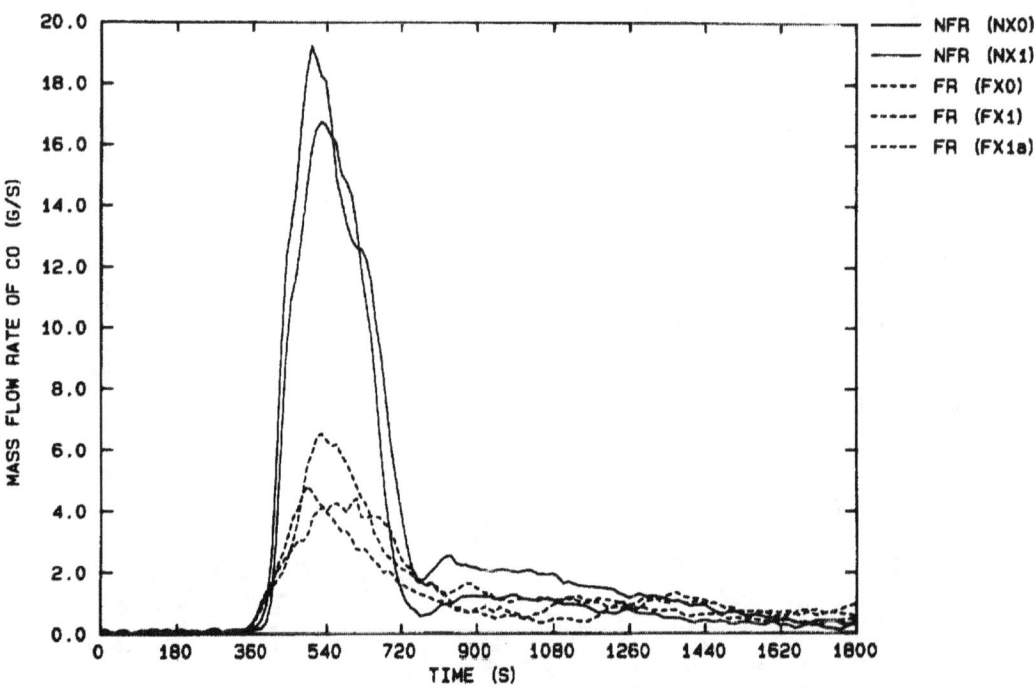

Figure 49. Mass flow rate of CO in the large-scale room/corridor/room tests with the auxiliary burner.

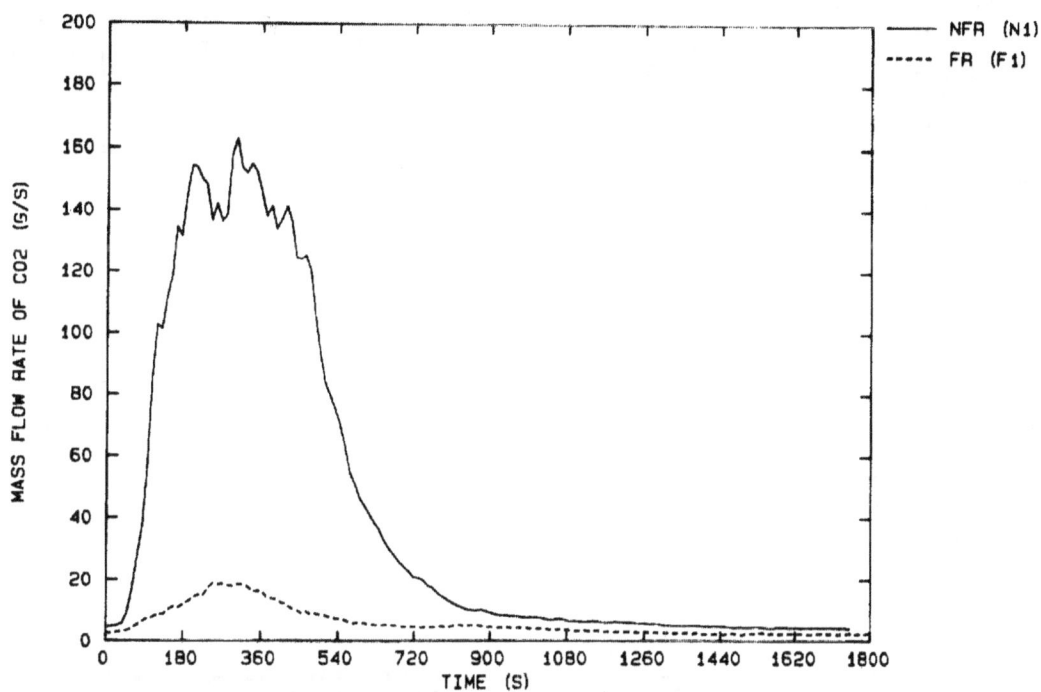

Figure 50. Mass flow rate of CO_2 in the large-scale room/corridor/room tests without the auxiliary burner.

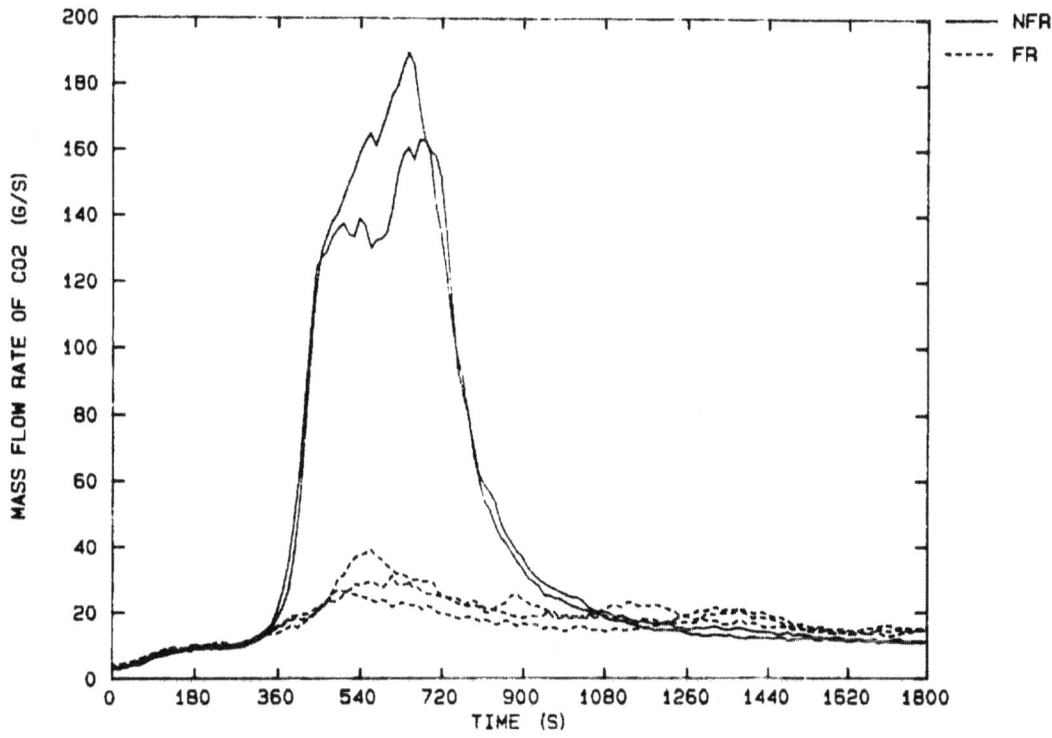

Figure 51. Mass flow rate of CO_2 in the large-scale room/corridor/room tests with the auxiliary burner.

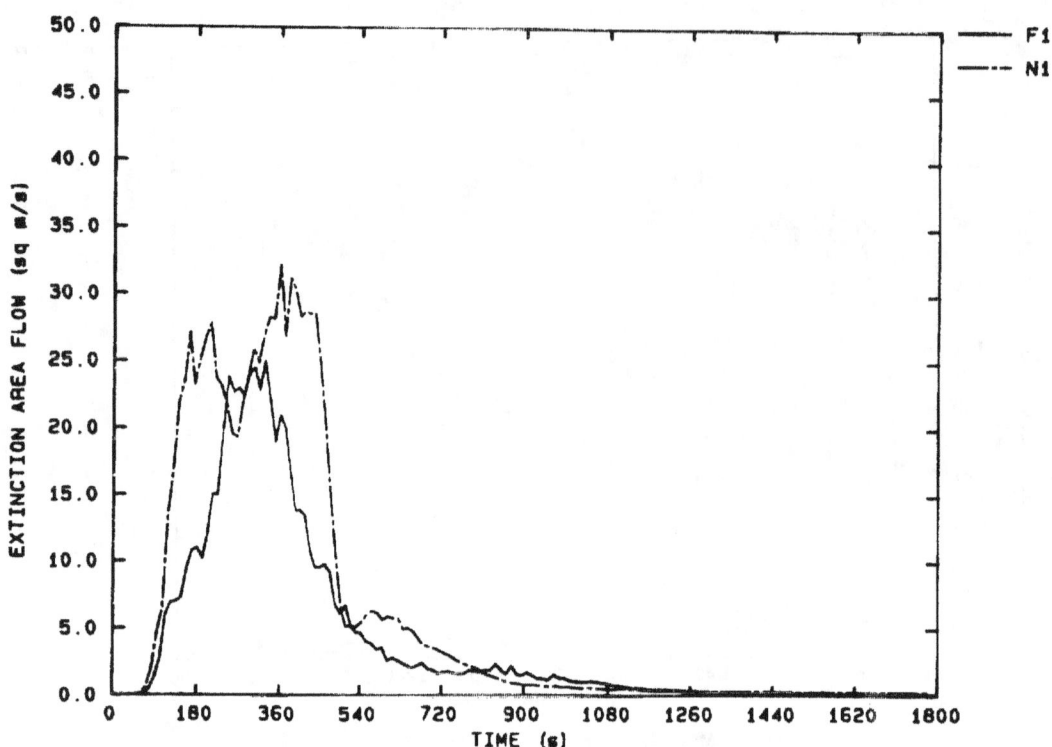

Figure 52. Smoke flow rates in the large-scale room/corridor/room tests without the auxiliary burner.

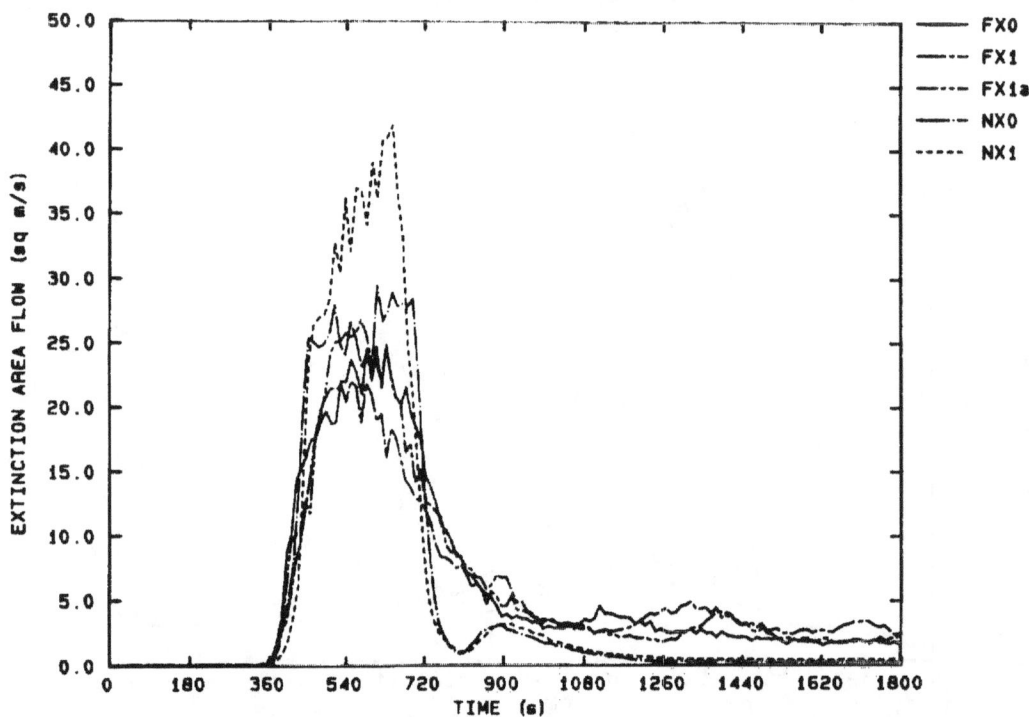

Figure 53. Smoke flow rates in the large-scale room/corridor/room tests with the auxiliary burner.

6

Assessment of Hazard

THE analysis of the fire hazard can be considered two ways. Within the room of fire origin (or within a nearby room), the time available from the start of the fire until the time that untenable conditions are reached is the crucial issue. On the other hand, if the fire is occurring in a large building, where significant distribution of the products of combustion is likely to occur via flows through ventilating systems, stairwells, etc., then the exact time of incapacitation cannot be derived for a far-removed occupant except by doing detailed flow calculations. In such a case, however, it is possible to postulate that the hazard will be proportional to the total *amount* of the products of combustion (heat, smoke, toxic gases) that is released from the fire. For completeness, both analyses will be made.

6.1 TIMES TO UNTENABILITY

6.1.1 FLASHOVER

The ultimate event defining when a room becomes totally untenable for the occupants is the occurrence of flashover. Flashover is essentially the point at which the room becomes filled with flames. Quantitatively, it can be defined in various ways. For the purpose of evaluating the present large-scale results, it will be taken as the time when 600°C is reached in the upper gas layer of the room. The temperature values will refer to two 0.05 mm thermocouples located near thermocouple tree 1 and 0.10 m down from the ceiling in the burn room. These values are listed in Table 23 for the burn room (no flashover events occurred in the target room in any of the tests). The difference between the performance of the NFR and the FR series is clear—in the NFR tests flashover occurred at just under two minutes, at 113 s, on the average. For the FR tests, meanwhile, flashover never occurred. For those FR tests, the highest temperature reached was 334°C, which is substantially less than the 600°C for flashover.

6.1.2 HEAT FLUX

If the heat flux is excessive during a room fire, occupants may be receiving such painful heating that viable escape options are not open. A heat flux value of 2.5 kW/m² is normally used as the criterion, determining when untenability due to excessive heat flux will occur [5]. For those tests where the 120 kW auxiliary burner was used, the criterion value was exceeded from the burner heating alone. Thus, those tests could not be evaluated under this criterion. For the remaining two tests, for the NFR products (test N1), the heat flux limit was exceeded in 41 s. For the corresponding FR test (test F1), the limit was exceeded at 73 s, which is a significant improvement. These results should not be used as the sole comparison of the relative hazard, however, since the test without the auxiliary burner does not represent a suitably high challenge to the FR products.

6.1.3 FRACTIONAL EFFECTIVE EXPOSURE-DOSE

Another variable which can have a strong effect on tenability is the toxicity of the combustion gases. Since the gas concentrations in the large-scale tests (as opposed to the NBS Combustion Toxicity Test) are not constant over time, the time factor has to be explicitly accounted for. This is most appropriately expressed as a "fractional effective exposure-dose" (FED). When all the toxic gas components are being determined at each time step, it is possible to use simple summation expressions to evaluate the FED at each time step [5].

For the present test series, the measurement of certain gases (HCN, HCl, HBr) could not be done on a time-resolved basis; those gas data are available only as over-all values for the entire test period. Thus, an analysis approach was developed which is based on the [CO toxicity/total toxicity] fractions computed above. By this approach, if [CO toxicity/total toxicity] is equal to, for example, 0.3, then the FED contribution *for CO alone* is summed at the appropriate location — burn room or target room — until FED = 0.3 is reached. At that instant of time, toxicologically untenable conditions are presumed to occur. The CO contribution to the FED was computed according to the method given in [5]:

$$= \frac{\sum_i (CO_i - 1700) \Delta t_i}{80,000} \quad (6)$$

Using the CO toxicity ratios given in Table 22, toxic hazard data based on the FED of CO are also presented in Table 23.

Excluding the results from test F1, the results show that for the NFR tests the critical CO FED was reached in an average of 166 s in the burn room and 214 s in the target room. By contrast, the FR tests showed that it took an average of 1789 s to reach the untenable value in the burn room. In the target room, the untenable value was never reached, except for test FX1a, where it was reached in 1013 s. Thus, the escape time from the burn room, evaluated on a toxicity basis, was increased by over 10-fold for the FR configuration, as compared to the NFR control.

6.1.4 SMOKE

In some large-scale fire testing a third criterion, time to reach a prescribed vision obscuration due to smoke, is included. Since in the present test series no instrumentation was included in the burn room for measuring smoke, therefore, this criterion is not applied. Note, however, that smoke production data (below) are available and are important in evaluating the product performance.

6.1.5 OVERALL UNTENABILITY

As stated above, the data from the heat flux meter should not be used in rating the actual performance. This leaves two main criteria—flashover and the CO FED. The *earlier* of the two governs the amount of escape time available for occupants. Thus, for the NFR tests, this is the 113 s until the reaching of flashover in the burn room. For the FR tests, the time of untenability occurs at 1789 s, when the critical CO FED is first reached. Thus, the occupants of the burn room would have more than 15-fold the escape time available for the FR case than for the NFR.

The comparative results in the target room were very similar (Table 23). However, since no flashover was achieved there for any of the tests, and for most FR tests the CO FED critical value was never reached, a quantitative evaluation is not made.

6.2 AMOUNTS OF COMBUSTION PRODUCTS

6.2.1 MASS CONSUMED

One beneficial effect of a fire retardant system is typically to reduce the amount of specimen mass burned. Table 19 presents the data on specimen mass lost during the large-scale tests. In the case of the NFR tests, it was assumed that all consumable weight loss occurred during the 30 or 35 minute test. Thus, these specimen masses refer to the mass loss as recorded for each item tested in the Furniture Calorimeter (Table 7). For the FR tests the only two items involved in the fire were the television cabinet and the chair. The cabinet was visually observed to be consumed during the 30 or 35 minute tests; thus, it was taken to be 100% consumed in computations. The amount of chair consumed after 30 or 35 minutes was estimated by observing the video tapes of the fire tests and these are the mass loss values recorded in Table 19. Certain large-scale results can be averaged for this comparison. The presence or absence of animals certainly does not affect specimen combustion. For the NFR case, it can be concluded that the presence or absence of the auxiliary burner had negligible effect; thus, the N and the NX tests can be averaged together also. Excluding test F1, then, the average mass lost for the FR tests was 12 kg, compared to 30 kg for the NFR case. This more than two-fold improvement is significant.

6.2.2 HEAT RELEASED

While the Cone Calorimeter, Furniture Calorimeter, and bench-scale toxicity protocol tests help in identifying the behavior of materials, the final evaluation has to rest with the large-scale result. One simple comparison that can readily be made is the total heat release measured from the large scale tests. Table 24 presents these data. The total heat release values for the individual large-scale tests were taken from the computer output at 100 s after the completion of the 30 or 35 minute tests. The tests where NFR materials were used showed an average of 542 MJ released, while the corresponding FR tests showed only 121 MJ. However, as can be seen from Figure 42, the fires have an extended smolder period after active burning is over. The rates of heat release occurring after the peak burning is passed were small, on the order of 100 kW. Since it can be assumed that there is an uncertainty of approximately ±50 kW associated with the measurement, it can readily be seen that when the heat release rate curve is integrated, a substantial error can occur if a significant amount of burning occurred during the smoldering period. Thus, it is appropriate to also consider the data from the Furniture Calorimeter and the Cone Calorimeter in attempting to arrive at the best estimate of the actual heat released. The values reported in Table 24 have been derived by using the specimen mass losses recorded in Table 19 and the values of Δh_c given in Table 8. For the NFR products, the Furniture Calorimeter and the Cone Calorimeter data agree very closely, indicating approximately 750 MJ. For the FR products, the heat of combustion for one of the main fire-involved items, the chair, could not be determined in the Furniture Calorimeter, due to the lack of flame spread in that test. Since the data for the two devices agree so closely for the NFR case, however, it is appropriate to consider that the value of 200 MJ derived from Cone Calorimeter measurements for the FR case is the correct one.

The "true" heats released are, then, taken as 750 MJ for the NFR case and 200 MJ for the FR case, with an improvement of 3.5 to 4-fold over the NFR case.

Using the data on the specimen mass burned above, effective heats of combustion can also be computed for the large-scale tests. These are:

25.0 MJ/kg for the NFR case
16.7 MJ/kg for the FR case.

Thus, it is seen that the improvement in the heat released for the FR case is a combined effect, due in part to the decreased specimen mass consumed and, in part, due to the lowered heats of combustion.

6.2.3 PRODUCTION OF SMOKE

Table 25 gives the total *production* of smoke, measured for the duration of the entire test. This was determined by integrating the values recorded by the smoke photometer located in the exhaust collection system. The NFR tests showed an average of 9900 ± 1000 m², while the FR tests showed 12400 ± 500 m². These are not statistically significantly different values at 2σ.

Table 25 also lists the *yields* (per-mass-burned values) of smoke in the large-scale tests, and compares the yields that would have been predicted on the basis of Furniture Calorimeter and Cone Calorimeter data. Both these latter tests over-predict the yield for the NFR tests and under-predict it for the FR ones.

From the Furniture Calorimeter data in Table 8, it was concluded that the smoke *yields* were very similar for NFR as for FR products, with the sole exception of the polystyrene TV cabinets. The large-scale findings, however, indicate a similar *amount* of smoke emitted, even though the amount of product burned in the FR case was only about 1/3 of that for the NFR case. Part of the explanation for this lies in the significant role of the polystyrene TV cabinets. The TV cabinets, both NFR and FR, are the smokiest product in the test series, and are about 10 times as smoky as the chairs. For the FR tests, even though the consumed mass from the TV cabinet amounts to only 18% of the total, according to the Furniture Calorimeter data (Table 8), the cabinet contributes

$$2800 \times 2.1(2800 \times 2.1 + 180 \times 9.9) = 77\%$$

of the total smoke production. Thus, since the FR TV cabinet is smokier than the NFR one, it is unsurprising that the total smoke production is not diminished for the case of the FR tests.

6.2.4 PRODUCTION OF CO

Table 26 gives data on the production of CO. The total yield of CO (kg) was determined from the gas analysis measurements made in the exhaust collection system. Averaging the tests in groups, as before, the average large-scale production of CO is, then, equal to 5.45 kg for the NFR case and 2.77 kg for the FR case. The hazard associated with CO production, therefore, is decreased by a factor of 2 for the FR case.

Also listed in Table 26 are the yields, computed on a basis of kg CO/kg fuel mass lost. There is no significant difference between the 0.18 kg/kg for the NFR case, versus 0.23 kg/kg for the FR tests. Comparing now these large-scale results to the data from the Furniture Calorimeter and the Cone Calorimeter, it can be seen that the predictability is generally poor. The CO yield in the Furniture Calorimeter, as discussed above, is substantially higher than in the Cone Calorimeter. In turn, the CO yield in the actual large-scale fire is twice again what was predicted from the Furniture Calorimeter data.

6.2.5 TOTAL PRODUCTION OF TOXIC GAS SPECIES

Measuring capability for HCN, HCl, and HBr was not available in the exhaust collection system. It is still possible to draw conclusions on the total toxic effluent which was collected by making use of supplementary bench-scale data. This evaluation can be accomplished in the following way. In the section above, the quantities of CO produced were itemized. To arrive at an assessment of the total toxicity, these values must be divided by a factor which represents [CO toxicity]/[total toxicity]. This is done very similarly as was done above in determining the time to reach the critical FED. Table 27 presents these results. These indicate an average value of 17.9 for the NFR products, compared to 5.7 for the FR ones. Therefore, the contribution to toxic hazard from the FR products is decreased 3-fold, compared to the NFR ones.

7

Summary

7.1 PROPERTIES OF INDIVIDUAL PRODUCTS TESTED

7.1.1 CONE CALORIMETER: IGNITABILITY

IGNITABILITY measurements in the Cone Calorimeter showed the following. None of the specimens, NFR or FR, was especially susceptible to ignitability, which is taken as showing ignition at an irradiance of 10 kW/m² or less. At an irradiance of 30 kW/m², specimen S showed an ignition time which was about double that of the NFR control, while specimen L showed an improvement of 50%. For the remaining specimens, the FR agent made essentially no difference. At the high irradiance of 100 kW/m², the FR agent did not change ignitability significantly for any of the specimens.

7.1.2 CONE CALORIMETER: PEAK RATE OF HEAT RELEASE

In almost all cases, the FR product showed a 2- to 3-fold improvement over the NFR one. The one exception was the electrical cable, when tested at 30 kW/m², where no difference was seen.

7.1.3 CONE CALORIMETER: SMOKE YIELD

Smoke yield, per-kg-of-specimen-mass-loss, was generally the same for the FR specimens as for the NFR ones. The one exception was the polystyrene TV cabinets, where the FR specimen showed about twice the smoke yield as did the NFR one. The hazard assessment, however, is not just based on this yield data, since it comes from total production values, not simply on the value for the smoke yield. This requires knowing what the actual mass loss rates will be for the large-scale test article, and cannot be obtained from Cone Calorimeter data alone, unless a predictive relationship has already been determined.

7.1.4 CONE CALORIMETER: CO YIELD

The *yields* of CO from the FR products were typically higher than from the NFR ones. This does not necessarily translate into higher *production* of CO for the FR cases, since the amount burned has to be also considered to determine that. The amount of specimen burned in the large-scale cannot as yet, in most cases, be predicted from bench-scale measurements. Thus, it is more appropriate to consider the actual product toxicity hazards in the context of the large-scale findings.

Here, we just note, further, that for some NFR specimens the CO yields obtained in the Cone Calorimeter are low, in the range of 0.01 to 0.05. By contrast, measurements in the large-scale tests (and in the Furniture Calorimeter) suggest that values in the vicinity of 0.1 are to be expected.

7.1.5 FURNITURE CALORIMETER FINDINGS

In all cases, the FR product was notably better-behaved in its rate of heat release characteristics, compared to the NFR one. In some cases (the TV cabinets, the business machine housings, the Z-arrangement cable arrays, and the circuit boards) the FR product showed, very roughly, half the RHR of the un-retarded one. In other cases (the chairs and the cable arrays in the vertical arrangement) the FR product showed no continued flame propagation at all, leading to RHR values which are much less, but not strictly comparable to, those registered by the NFR product. The yields of CO and smoke, on a per-kg-of-specimen-burned bases, were not greatly different, except for the TV cabinets (CO and smoke) and the business machine housings (CO). In those cases, the production from the FR specimens was 2 to 3 times that from the NFR ones. However, since the burning rate was also reduced by 13 to 23, the net hazard associated with the CO or smoke production was not significantly increased over the NFR case.

Caution must be exercised in interpreting and using Furniture Calorimeter data when a no-continued-flame-spread result is reported. In the Furniture Calorimeter tests, neither the FR cables nor the FR chair showed sustained spread. As a result, for each the rate of heat release of the FR specimen was several-fold lower than for the NFR one. In the large-scale tests, however, the FR cables still showed no fire involvement, but the FR chair was consumed. The difference is attributed to different ignition sources. In the large-scale tests, a scenario was specified where the igniting burner was supplemented by the auxiliary burner. There was also more than one item in the room and, thus, there were interactions among burning items. The general caution, then, is that the results for the FR item should not be stated as being "x many times better" than for the NFR one if the FR one showed low heat release due to failure to spread flame over its entire surface.

7.1.6 SMALL-SCALE TOXICITY TESTS

A modified NBS Toxicity Test protocol was used, whereby a full LC_{50} determination was not made, but, rather, animal death results were compared against gas analysis predictions at a few specimen mass loadings. In no case was extreme toxic potency, identified as $LC_{50} < 2$ mg/l, observed. The biological results could be predicted on the basis of chemical analysis for CO, CO_2, low O_2, HCN, HCl, and HBr for all specimens except the FR polystyrene TV cabinet (specimen G), the FR cable wire insulation (specimen K), and the FR polyester circuit board (specimen L). For these three materials, additional agents may be needed to account for 20–40% of the observed toxicity. For the NFR chair (specimen T), this question could not be answered due to the experimental difficulties in quantifying very low toxicity foam specimens. NFR wire insulation (specimen D) may also need additional gases to account for no more than 20% of the toxicity.

7.2 RESULTS FROM LARGE-SCALE EVALUATIONS OF NFR- AND FR-FURNISHED ROOMS

7.2.1 IMPACT OF FR MATERIALS ON THE SURVIVABILITY OF OCCUPANTS OF ROOM OF FIRE ORIGIN

This is judged by the time that is available to the occupants before the earlier of (1) room flashover, or (2) untenability due to toxic gas production occurs. For the NFR tests, the average available escape time was 113 s. By contrast, for the FR tests, the time of untenability did not occur until 1789 s. Thus, the occupants of the burn room would have more than 15-fold the escape time available for the FR case than for the NFR.

7.2.2 AMOUNT OF MATERIAL CONSUMED IN FIRE

The FR tests showed less than half the amount of specimen mass lost as was lost in the NFR tests. Since mass loss is necessary before any hazardous combustion products can be generated, it can be concluded that there is an effective hazard control at the very outset.

7.2.3 AMOUNT OF HEAT RELEASED FROM THE FIRE

The FR tests indicated an amount of heat released from the fire which was 3.5 to 4 times less than that released by the NFR tests.

7.2.4 TOTAL PRODUCTION OF TOXIC GASES

The total quantities of toxic gases produced in the room fire tests, expressed in "CO equivalents," were reduced 3-fold by the use of the FR products, compared to the NFR ones.

7.2.5 PRODUCTION OF SMOKE

The production of smoke was not significantly different between the room fire tests using NFR products and those with FR products. The smoke production in the FR tests was dominated by smoke from the polystyrene TV cabinets. This is consistent with the smaller-scale tests, which showed smoke yield improvements for all FR products except for the TV cabinet. For the remaining four product categories, the production of smoke from the FR products was substantially smaller than from the NFR ones.

7.2.6 EFFECT OF AUXILIARY BURNER

The presence or absence of the auxiliary burner made no effect on any test results for NFR materials. For the FR tests, by contrast, the burner was clearly proven necessary to obtain non-trivial fire involvement. In the one FR test run without the auxiliary burner, the test chair was only partly consumed and the specimen mass loss was less than half that for tests where the auxiliary burner was used.

7.2.7 COMPARISON TO PREVIOUS STUDIES

The results of the present large-scale study contrast

sharply to some earlier studies on fire-retarded products [e.g., 2]. In that earlier study the FR products examined had a minimum retardant level, consistent with lowest possible cost and the need to pass only the simple California Bulletin 117 test [26]. Thus, not unexpectedly, the FR products in that study did not show measurably improved behavior. When FR formulations were chosen, as was done in the present study, to represent high-quality, rather than minimum cost systems, an entirely different, much more favorable result is seen.

7.3 RELATIONS BETWEEN MEASUREMENTS IN DIFFERENT TEST METHODS

7.3.1 PREDICTING HEAT RELEASE RESULTS FROM CONE CALORIMETER DATA

Can the Cone Calorimeter rate of heat release measurements of fire-retarded products indicate the level of improvement which the FR additive can affect in the large scale? In answering this question, for three of the five products tested, the bench-scale Cone Calorimeter measurements were seen to reflect very well the large-scale Furniture Calorimeter results. The FR products' improvement in performance was correctly predicted and to a roughly similar quantitative magnitude. In the remaining two cases, such was not the case. These were the electrical cables and the upholstered chairs.

The electrical cables presented a new issue in bench-scale testing. In almost all studies of heat release rate prior to this work, it had been found that when bench-scale specimens were prepared as composite, through-the-thickness samples cut out of the large-scale article, an adequate representation of the burning characteristics was obtained. During the course of this work, one of the first instances has been identified where bench-scale samples prepared in this way were not indicative of large-scale behavior. The electric cable samples, when tested as such composites, did not accurately reflect the contribution of the fire retardants located in the inner layers. The reason for this was seen to be that the gasified FR agents from the inner layers have an effect quite far away from the immediate area of combustion. They not only diminish burning in the vicinity of the pyrolysis region, but also prevent flame spread further up the cable array. This problem deserves further study, in order to derive suitable predictive capabilities.

The upholstered chair data from the Cone Calorimeter did not agree with the measurements taken in the Furniture Calorimeter. In the Furniture Calorimeter, the FR specimen performed well enough so that progressive flame spread did not occur, and only very little of the specimen was burned. The bench-scale data, while showing clearly the superiority of the FR formulation, did not indicate the possibility of such no-spread behavior. It is more important to note, however, that the data from the Cone Calorimeter did agree with the actual large-scale test results, where the upholstered chair did spread flame and burn nearly to completion. It cannot be overemphasized that while there may be wrong fire tests, in the sense of ones based on such faulty physics as to never be useful, there are not, in general, right fire tests—there can only be tests which are suited to a specific scenario, application, product category, etc.

7.3.2 SCALE EFFECT ON TOXIC AGENTS PRODUCED

7.3.2.1 Fraction of Total Toxicity Accounted for by CO

A comparison was made between the fraction of total toxicity accounted for by CO in the bench-scale NBS combustion toxicity test and the fraction measured in the large-scale room/corridor/room tests. For both NFR and FR materials, the large-scale results fell in the range circumscribed by the bench-scale data. This suggests that neither the differences in scale nor in the type of exposures involved in these two devices preferentially change the degree of toxicity produced by CO. The NBS combustion toxicity test is, thus, a reasonable device with which to characterize the CO.

7.3.2.2 Yields of the Toxic Gases

If the yields are similar for all species of interest in two different test devices, then it may be concluded that, at least for hazard assessment purposes, the essential aspects of combustion chemistry are preserved. In the present test series, CO_2 and HCl measurements showed very close agreement. Because of few measurements and high scatter, no firm conclusions for HBr or HCN yields are made.

For CO, there was substantive disagreement, with the NBS combustion toxicity apparatus and, especially, the Cone Calorimeter giving some values which were substantially lower than those observed in the Furniture Calorimeter.

The large-scale results showed that there was no difference in CO *yield* between the NFR and the FR cases. In other words, the CO *production* for NFR versus FR was in the same 2:1 ratio as were the respective mass losses. The predictability of these CO yields was generally poor. The CO yields in the Furniture Calorimeter were only half of what was noted in the large-scale tests. In turn, the CO yields in the Cone Calorimeter were several-fold lower yet than what was observed in the Furniture Calorimeter. This lack of scaling did not especially depend on whether the products were NFR or FR. The reasons for this lack of agreement can be manifold, however, it can be surmised that both flame residence time and turbulent mixing scale or intensity are governing chemical factors which are inseparably associated with scale. Indeed, the CO data from the Furniture Calorimeter showed little variation amongst the specimens, both NFR and FR.

The CO yields in the NBS Combustion Toxicity Test were lower than in the corresponding large-scale tests, while the yields of the remaining species being measured were generally much more similar across all test methods. Thus, for the question of whether *unusual* toxic species are present, since these bench-scale results are biased in favor of the non-CO species, they are expected to yield false positives but not false negatives. Such a conservative bound to the true results is acceptable for a screening test. For resolving the question of *extreme* toxicity, it is necessary to consider the behavior in the large-scale tests. The approximately 0.2 yield seen in the large-scale results, in fact, is near the upper limit to CO yields, as seen in other studies [27]. Thus, it is clear that extreme toxicity in large scale could not result by the increase of CO yields by a necessary order of magnitude or more. Instead, if any products show extreme toxicity in large scale, it has to be by the increased quantities of the other species. The NBS Combustion Toxicity test, however, as demonstrated above, is biased in favor of the non-CO species. Thus, also for indicating extreme toxicity it can be presumed to be a method where false positives, but not false negatives, can be expected, and is, therefore, conservative.

For accurate fire modeling to be achievable, it will be necessary to predict correctly the CO in large-scale fires. The data gathered in the present study reinforce the growing evidence that the production of CO in large-scale fire tests (and, thus, presumably in real fires) is only somewhat influenced by the chemical properties of the substance being burned. The CO production is much more influenced by the availability of oxygen in the large scale fire. This, in turn, is affected by variables such as geometry, ventilation configuration, turbulence, and mixing. If this view is correct, the usage of any less-than-room-sized tests for making CO predictions has to be deferred until these oxygen supply variables are sorted out. Thus, in the near future, emphasis has to be placed on the development of predictive algorithms which take into account fluid mechanics, gas phase reactions, multiple-item interactions, and other full-scale phenomena.

7.3.3 PREDICTING SMOKE YIELD RESULTS FROM CONE CALORIMETER DATA

In all cases, the ratio between the smoke yields for the NFR and the FR product were similar in the Cone Calorimeter and in the Furniture Calorimeter.

The relationship between the numerical values in the Furniture Calorimeter, versus those in the Cone Calorimeter was typically 1:2. The exception to this diminished smoke production in the larger-scale test came from the polystyrene TV cabinets. For the polystyrene specimens, both the NFR and the FR versions showed a greater smoke yield in the Furniture Calorimeter than in the Cone Calorimeter.

7.3.4 USE OF LESS-THAN-ROOM-SCALE TESTS TO PREDICT ROOM FIRE BEHAVIOR

Prediction of room fire behavior from smaller-scale test data is a relatively recent area of fire research. A basic prediction would require that both the burning rate (heat release rate or mass loss rate) and the relevant yields of products could be predicted. The yields have already been discussed. Attempts to predict the burning rates are a new, yet important, area of endeavor. There are, generally, a number of physical phenomena which cannot be adequately represented by the smaller-scale test itself, and must be accounted for by special data analysis techniques or by empirical correction factors. Nonetheless, since there is a huge economic incentive to utilize small-scale test wherever possible, further advances need to be made. The techniques to predict the rate of heat release will be highly dependent on the type of commodity being considered. Of the products being evaluated by FRCA, only for upholstered furniture has work already been done on relationships to predict the large-scale rate of heat release from Cone Calorimeter data. For design purposes, the predictability for NFR items is probably satisfactory, however, further work would still need to be done on FR items, to determine, e.g., the threshold value from the Cone Calorimeter below which continued flame spread is to not be expected. For the remaining classes of commodities, pursuit of such predictive relationships is still work for the future.

Conclusion

THE fire performance of fire-retarded and non-fire-retarded versions of several product types have been studied using state-of-the-art fire science. It has been demonstrated for representative fire-retarded products that significantly enhanced fire performance can be obtained, in that:

- The average available escape time was more than 15-fold greater for the FR products in the room burn tests.
- The amount of material consumed in tests of the FR products was less than half the loss in the NFR tests.
- FR products, on the average, gave 1/4 the heat release of NFR products.
- The total quantities of toxic gas, expressed as CO-equivalents, released by the FR products was 1/3 of that for the NFR products.
- The production of smoke was not significantly different in room fire tests between FR and NFR products.

The study shows, then, that the proper selection of fire retardants can improve fire and life safety.

The conclusions developed in this study are pertinent only to the materials actually examined. Such improvements are not to be automatically expected from all fire-retarded products. Instead, it will still be necessary to test and evaluate proposed new systems individually. The research program provides a methodology to do that.

The new instruments and analyses used to predict the performance of these materials show promise. The main current limitation is the inability to predict accurately the production of CO from less-than-room-sized tests. Specific research efforts will have to be addressed to solve this issue. Also, long-term efforts are needed to develop specific methods for predicting the detailed burning rate characteristics of different classes of commodities.

APPENDIX A

Ion Chromatography Procedure

APPARATUS

A commercially available ion chromatograph (Waters Model ILC-1 IonLiquid Chromatograph) equipped with a Waters 430 Total Conductivity Detector and a Waters 460 Electrochemical Detector was used to analyze for Br^-, Cl^-, and CN^-. The electrochemical detector was used with an Ag working electrode and a saturated KCl reference electrode. An anion column (ICPAK-A) preceded by an Anion Guard-Pak Precolumn Module, both commercially available from Waters, was used. Chromatograms were recorded on a Spectra-Physics Model SP 4270 Integrator.

REAGENTS

All chemicals used in this work were of reagent grade quality. The water used was conditioned to 18.3 MΩ-cm and passed through a 0.45 μm nominal porosity filter. The eluent for the ion chromatograph was 5 mM KOH. Stock solutions of Br^-, Cl^-, and CN^-, nominally 1000 ppm, were prepared by dissolving 0.1489 g of KBr, 0.2100 g of KCl, and 0.2502 g of KCN, respectively, in 100 mL of the eluent described above. Calibration solutions of 1.0 to 5.0 ppm for Br^- and Cl^- and 0.01 to 0.03 ppm for CN^- were prepared by serial dilution of the stock solutions with the eluent.

CHROMATOGRAPHIC PROCEDURE

The eluent flow rate through the system was 1.0 to 1.2 ml/min. The sample loop had a volume of 100 μl. Unknowns were diluted 1:10 for Br^- and Cl^- and 1:100 for CN^- with eluent. (There were a few unknowns that required a 1:1000 or 1:500 dilution for CN^-.) Samples and standards were loaded into the loop using a syringe and a 0.45 μm syringe filter. The sample loop was rinsed with ca. 1 ml of the analyte solution before the sample was injected onto the column. The procedure, as evolved and described above, was successful in minimizing any mutual interferences among the three anions of interest.

APPENDIX B

Gas Chromatography Procedure

THE analytical procedure for the HCN determination was based on a gas chromatographic (GC) technique utilizing an alkali flame (or thermionic) detector. The details of this technique have been published [20].

A commercially available gas Chromatograph (Hewlett Packard 5880) was used. Briefly, the analysis was carried out on a 1.8 m × 0.32 cm OD (6 ft. × 1/8 in. OD) stainless steel column packed with Porapak Q, 80/100 mesh. Column temperature was maintained at 110 C, injection port at 200 C, and the detector at 300 C. Helium was used as the carrier gas at 30 ml/min. Air at 120 ml/min. and hydrogen at 3 ml/min. were used as the detector gases. The combustion products were injected with a gas-tight syringe. Quantitation was based on integrated peak areas.

Calibration of the alkali flame detector was performed with commercially supplied mixtures of HCN in nitrogen. The concentration of HCN gas mixtures was verified by titration with standard silver nitrate solution [25].

APPENDIX C

Log of Large-Scale Test Observations

Test F1

Time (s)	Comment
60	side of couch flaming
120	top side aflame very heavy smoke
200	burner off
230	still flame above TV
283	flames almost gone
315	flames out
420	target room peak CO in bot. chamber, 3800 ppm burn room peak temp 220°C
660	inside of chair flaming (next to TV)
1140	flames out
1800	test over; auxiliary burner off

Post-fire analysis:

T.V. cabinet – consumed
left cushion – consumed
back cushion – 1/2 consumed
seat cushion – 1/4 consumed
right cushion – no damage

Test FX0

Time (s)	Comment
0	120 kW auxiliary burner on
300	50 kW burner on
325	cabinet burning
330	couch
390	1/3 TV cabinet gone top left arm couch aflame
500	50 kW burner off
510	temp in burn room same as before
810	flames covering almost entire top of seat cushion
900	circuit board on left burning
960	right side of cushion aflame
1770	back cushion fell on seat
2010	top of right cushion arm aflame
2100	test over; auxiliary burner off
2700	data logger off

first 6 inches backside cables intact
next 20 inches badly charred white, yellowish ash, copper strands exposed
next 42 inches covering split on most of cables exposing insulation
business machine cabinets were warped and shrunken
1/4" circuit board charred to depth of 1/16 to 1/8" only in regions behind T.V. cabinets & chair "S"

Test FX1

Time (s)	Comment
0	120 kW auxiliary burner on ($\Delta P = 0.84$)
0+	flames on ceiling
300	50 kW ignition turned on
330	flame collapsing back of TV cab
360	back TV consumed, only 1/5 TV frame consumed
390	all back TV and 1/2 front TV gone
420	both almost gone, arm of chair aflame, heavy dense smoke from corridor
500	50 kW off burner
540	melt from TV burning on floor
780	seat cushion aflame for about 6 in. from left arm.
900	lower 6 in. height of back cushion aflame and 18 in. of flames across seat cushion
960	flames reached across only to right arm (which has not ignited) flames about 18 in. tall on seat cushion also burning on top of left arm of chair
1170	inside right arm aflame along entire height
1320	top of right arm burning vigorously
1410	fire penetration of right arm cushion only near top
1729	back cushion fell on seat and burning vigorously

Time (s)	Comment
2100	auxiliary. burner turned off
2280	right cushion penetrated, cables *may* be involved
2400	impinger sampling completed
2520	test data over

Test N1

Time (s)	Comment
0	50 kW burner on
30	cushion inside involved
120	flashover (600°C)
170	fan to boxes turned on
200	burner off
600	smoke clearing up; chair gutted; circuit board burning but intact; cables gone
660	only business machine burning
840	only remnants of business machine left and burning; circuit board intact
1020	only 2 flamelets left on business machine
1920	test over

Test FX1a

Time (s)	Comment
0	120 kW auxiliary burner on
300	50 kW ignition source on
315	TV penetration on bottom
325	couch ignited
390	heavy smoke; top left arm couch flaming
500	50 kW burner off
840	inside cushions of chair (all 4) aflame
1350	back cushion fell on seat
2085	only flamelets left on right side of chair
2100	test over; aux. burner off; only remnants of cushion left on seat and lower right side
3000	data logger off

Test NX0

Time (s)	Comment
0	120 kW auxiliary burner on
300	50 kW ignition source on
315	TV ignited
320	couch ignited
360	smoke 1/2 way down doorway
375	dense smoke
390	seat involvement
405	flashover chair
420	room flashover (~600°C)
500	50 kW burner off
720	fire clears; cables consumed
840	only business machine burning
2100	test over

Test NX1

Time (s)	Comment
0	120 kW auxiliary burner on
300	50 kW burner on
315	TV ignites; chair ignites
335	top right arm of chair ignites
390	flames half way to top of chair
420	flashover (600 'C)
435	smoke obscured fire completely
500	ignition burner off
697	fire clearing up; fire spread beyond chair
720	pump malfunctioned (maybe corridor room)
780	just business machine burning; cables completely burned up
1080	just flamelets left from business machine; circuit board folded up, partially burned
2100	test over
2970	data logger off

References

1. Babrauskas, V., "Development of the Cone Calorimeter—A Bench-Scale Heat Release Rate Apparatus Based on Oxygen Consumption" (NBSIR 82-2611). [U.S.] Nat. Bur. Stand. (1982).

2. Babrauskas, V., J. R. Lawson, W. D. Walton and W. H. Twilley, "Upholstered Furniture Heat Release Rates Measured with a Furniture Calorimeter" (NBSIR 82-2604) Nat. Bur. Stand. (1982).

3. Babrauskas, V., B. C. Levin, and R. G. Gann, "A New Approach to Fire Toxicity Data for Hazard Evaluation." *ASTM Standardization News*. 14, 28-33 (1986).

4. Braun, E., B. C. Levin, M. Paabo, J. Gurman, T. Holt and J. S. Steel, "Fire Toxicity Scaling" (NBSIR 87-3510). [U.S.] Nat. Bur. Stand. (1987)

5. Bukowski, R. W., W. W. Jones, B. M. Levin, C. L. Forney, S. W. Stiefel, V. Babrauskas, E. Braun, and A. J. Fowell, Hazard I. Volume I: Fire Hazard Assessment Method (NBSIR 87-3602). [U.S.] Nat Bur. Stand. (1987).

6. Levin, B. C., M. Paabo, M. L. Fultz, C. Bailey, W. Yin and S. E. Harris, "An Acute Inhalation Toxicological Evaluation of Combustion Products from Fire Retarded and Non-Fire Retarded Flexible Polyurethane Foam and Polyester" (NBSIR 83-2791). [U.S.] Nat. Bur. Stand. (1983).

7. Huggett, C., "Estimation of Rate of Heat Release by Means of Oxygen Consumption Measurements," *Fire and Materials*, 4, 61-65 (1980).

8. Babrauskas, V., and G. Mulholland, "Smoke and Soot Data Determinations in the Cone Calorimeter," pp. 83-104 in *Mathematical Modeling of Fires* (ASTM STP 983), American Society for Testing and Materials, Philadelphia (1987)

9. Babrauskas, V., "The Cone Calorimeter—A Versatile Bench-Scale Tool for the Evaluation of Fire Properties," pp. 78-87 in *New Technology to Reduce Fire Losses & Costs*, S.J. Grayson and D.A. Smith, eds., Elsevier Applied Science Publishers, London (1986).

10. Proposed Test Method for Heat and Visible Smoke Release Rates for Materials and Products using an Oxygen Consumption Calorimeter (P 190). *Annual Book of ASTM Standards*, Vol. 04.07, American Society for Testing and Materials (1986).

11. Levin, B. C., M. Paabo, M. L. Fultz and C. S. Bailey, "Generation of Hydrogen Cyanide from Flexible Polyurethane Foam Decomposed under Different Combustion Conditions," *Fire and Materials* 9, 125-134 (1985).

12. Babrauskas, V., "Will the Second Item Ignite?" *Fire Safety J.* 4, 281-292 (1981/82).

13. Babrauskas, V., "Upholstered Furniture Room Fires—Measurements, Comparison with Furniture Calorimeter Data, and Flashover Predictions," *J. of Fire Sciences*, 2 5-19 (1984).

14. Mulholland, G. M., V. Henzel Babrauskas, V., "The Effect of Scale on Smoke Emission," paper to be presented at Fire Safety Science, Second International Symposium (1988).

15. Babrauskas, V., "Free Burning Fires," *Fire Safety J.*, 11, 33-51 (1986).

16. Babrauskas, V. and W. D. Walton, "A Simplified Characterization for Upholstered Furniture Heat Release Rates," *Fire Safety J.*, 11, 181-192 (1986).

17. Quintiere, J. G., paper presented at the ASTM E-5 Research Review, Bal Harbour, FL (December 1987).

18. Levin, B. C., M. Paabo, J. L. Gurman and S. E. Harris, "Effects of Exposure to Single or Multiple Combinations of the Predominant Toxic Gases and Low Oxygen Atmospheres Produced in Fires," *Fund. Appl. Tox.*, 9, 236-250 (1987).

19. Levin, B. C., A. J. Fowell, M. M. Birky, M. Paabo, A. Stolte and D. Malek, "Further Development of a Test Method for the Assessment of the Acute Inhalation Toxicity of Combustion Products (NBSIR 82-2532)," [U.S.] Nat. Bur. Stand. (1982).

20. Paabo, M., M. M. Birky and S. E. Womble, "Analysis of Hydrogen Cyanide in Fire Environments," *J. Comb. Tox.*, 6, 99-108 (1979).

21. Jones, W. W., "Future Directions for Modeling the Spread of Fire, Smoke, and Toxic Gases," *Fire Safety: Science and Engineering*, ASTM STP 882, T. Z. Harmathy, Ed., American Society for Testing and Materials, Philadelphia, pp. 70-96 (1985).

22. Lee, B. T., "Quarter-Scale Modeling of Room Fire Tests of Interior Finish (NBSIR 81-2453)," [U.S.] Nat. Bur. Stand. (1982).

23. Steckler, K. D., J. G. Quintiere and W. J. Rinkinen, "Flow

Induced by Fire in a Compartment (NBSIR 82-2520)," [U.S.] Nat. Bur. Stand. (1982).

24 Jones, W. W., and R. D. Peacock, "Experimental Verification of a Model for Fire Growth and Smoke Transport," paper to be presented at the Second Annual Symposium on Fire Safety Science, Tokyo, Japan (June 1988).

25 Kolthoff, I. M. and E. B. Sandell, *Textbook of Quantitative Organic Analysis*, 2nd Ed., p. 546. The MacMillan Co., New York, (1953).

26 Flammability Information Package (Contains Technical Bulletins 116, 117, 121, and 133). Bureau of Home Furnishings, Dept. of Consumer Affairs, State of California, North Highlands (1985).

27 Mulholland, G. W., private communication (1988).

FORM NBS-114A (REV.11-84)

U.S. DEPT. OF COMM. BIBLIOGRAPHIC DATA SHEET (See instructions)	1. PUBLICATION OR REPORT NO. NBS/SP-749	2. Performing Organ. Report No.	3. Publication Date July 1988
4. TITLE AND SUBTITLE FIRE HAZARD COMPARISON OF FIRE-RETARDED AND NON-FIRE-RETARDED PRODUCTS			
5. AUTHOR(S) Vytenis Babrauskas, Richard H. Harris, Jr., Richard G. Gann, Barbara C. Levin, Billy T. Lee, Richard D. Peacock, Maya Paabo, William Twilley, Margaret F. Yoklavich, Helene M. Clark			
6. PERFORMING ORGANIZATION (If joint or other than NBS, see instructions) NATIONAL BUREAU OF STANDARDS U.S. DEPARTMENT OF COMMERCE GAITHERSBURG, MD 20899		7. Contract/Grant No.	
			8. Type of Report & Period Covered Final
9. SPONSORING ORGANIZATION NAME AND COMPLETE ADDRESS (Street, City, State, ZIP) Fire Retardant Chemicals Association Lancaster, PA 17604			

10. SUPPLEMENTARY NOTES

Library of Congress Catalog Card Number: 88-600560

☐ Document describes a computer program; SF-185, FIPS Software Summary, is attached.

11. ABSTRACT (A 200-word or less factual summary of most significant information. If document includes a significant bibliography or literature survey, mention it here)

A test program was conducted for the Fire Retardant Chemicals Association to quantify the effects of fire retardant chemicals on total fire hazard. Five different types of products, each made from a different type of plastic were used. The products were made up in analogous fire-retardant (FR) and non-retarded variants (NFR). Cone Calorimeter, Furniture Calorimeter, and NBS combustion toxicity tests were run on the individual products. The ultimate evaluation was made through the use of a large-scale burn room/corridor/target room facility. The test results showed very substantial improvements for the FR products in the areas of increased occupant escape time and decreased production of heat and toxic gas species. The production of smoke was unchanged with the FR products, compared to the NFR ones. It was specifically demonstrated that the overall hazard was reduced, and that reduced burning rates were not obtained at the expense of increasing the hazard due to combustion toxicity. Most of the combustion product effects could be accounted for by the normal gases monitored, and it was also shown that none of the products, either FR or NFR, constituted hazards due to extreme toxic potency.

12. KEY WORDS (Six to twelve entries; alphabetical order; capitalize only proper names; and separate key words by semicolons)

Cone Calorimeter; fire gas toxicity; fire retardant chemicals; Furniture Calorimeter; ion chromatography; plastics; rate of heat release; room fire tests; smoke production.

13. AVAILABILITY ☒ Unlimited ☐ For Official Distribution. Do Not Release to NTIS ☒ Order From Superintendent of Documents, U.S. Government Printing Office, Washington, DC 20402. ☐ Order From National Technical Information Service (NTIS), Springfield, VA 22161	14. NO. OF PRINTED PAGES 92
	15. Price

www.ingramcontent.com/pod-product-compliance
Lightning Source LLC
Chambersburg PA
CBHW080307180526
45167CB00006B/2700